VISUALISING UNCERTAINTY
A SHORT INTRODUCTION

A PUBLICATION OF THE
ANALYSIS UNDER UNCERTAINTY for DECISION MAKERS NETWORK

Written by
Polina Levontin and Jo Lindsay Walton

With additional contributions from
Lisa Aufegger, Martine J. Barons, Edward Barons, Simon French,
Jeremie Houssineau, Jana Kleineberg, Marissa McBride, and Jim Q. Smith

Layout and design by
Jana Kleineberg | Kleineberg Illustration & Design

Published in 2020 by AU4DM, London, UK
© Sad Press
ISBN: 978-1-912802-05-0

This edition was compiled with the support of the Sussex Humanities Lab.
With special thanks to Mark Workman of AU4DM.

The **Analysis Under Uncertainty for Decision Makers Network** (AU4DM) is a
community of researchers and professionals from policy, academia, and industry, who
are seeking to develop a better understanding of decision-making to build capacity and
improve the way decisions are made across diverse sectors and domains. For further
details about the network, visit http://www.au4dmnetworks.co.uk/.

Recommended citation
Levontin, P., Walton, J.L., Kleineberg, J., Barons, M., French, S., Aufegger, L.,
McBride, M., Smith, J.Q., Barons, E., and Houssineau, J.
Visualising Uncertainty: A Short Introduction (London, UK: AU4DM, 2020).

SUSSEX HUMANITIES LAB

FUTURE COLLABORATIONS
AND CONTACT

- **Jana Kleineberg,**
  Kleineberg Illustration & Design
  http://www.kleineberg.co.uk/
  jkl@kleineberg.co.uk

- **Polina Levontin**
  polina.levontin02@imperial.ac.uk *or*
  levontin@hotmail.com

- **Jo Lindsay Walton**
  https://www.jolindsaywalton.com/
  j.c.walton@sussex.ac.uk

- **Or contact us here**
  http://www.seaplusplus.co.uk/
  http://au4dmnetworks.co.uk/

CONTENT
VISUALISING UNCERTAINTY

1 SUMMARY
VISUALISING UNCERTAINTY

We must distinguish the different 'Types of Uncertainty' that we believe to be present.

See chapter 5, 'Types of Uncertainty'

The basis of this short introduction is a review of the literature on communicating uncertainty, with a particular focus on visualising uncertainty. From our review, one thing emerges very clearly: **there is no 'optimal' format or framework for visualising uncertainty.** Instead, the implementation of visualisation techniques must be studied on a case-by-case basis, and supported by empirical testing.

Developing solutions for uncertainty visualisation thus requires interdisciplinary expertise. The effectiveness of different techniques is highly context-sensitive, and current understanding of how to differentiate relevant contextual factors remains patchy. For this reason, **communication formats should ideally be developed through close collaboration among researchers, designers, and end-users.** The building blocks brought together here provide a starting point for these kinds of dialogues.

In order to develop an uncertainty visualisation format for a case study, **we must distinguish the different types of uncertainty** that we believe to be present (see 'Types of Uncertainty'). It is important to make these distinctions as clearly and as early as possible. Definitions and understandings of uncertainty should also be regularly reviewed as the design and testing process unfolds.

Without reliable evaluation methods, there is the danger of developing dazzling, seductive visualisations that fail to deliver appropriate decision support, or even subvert decision-making by slowing it down and/or introducing biases. Self-reporting by users is not a reliable way to assess the effectiveness of a visualisation format, so **evaluation should be performed by objective and reproducible methodologies.**

Users of uncertainty information have diverse capacities and needs, and there is as yet no deep theory which formalises which differences

Key ideas

case-by-case basis	*review & testing*
development through collaboration	*reliable evaluation*
different types of uncertainty	*reproducible methodologies*

There is no 'optimal' format or framework for visualising uncertainty.

are relevant in a given case. This is important, because **how we visualise uncertainty is not easily separable from how we interpret and reason about that uncertainty.** The diverse identities of decision-makers—our various cultural, political, social, linguistic, institutional, and individual characteristics—shape our practices in dealing with uncertainty. This means that effective uncertainty visualisation must ideally do more than encode all the relevant information: it must also invite, reinforce, and sometimes even teach appropriate modes of interpretation and reasoning. This catalogue offers illustrative discussion of selected heuristics and biases, and how they can interact with visualisation techniques. This gives a flavour of a vast area of research, and helps to illuminate some key features of the existing evidence base.

Although there are many challenges associated with visualising uncertainty for decision-making, there are also many potential benefits. In fact, the horizons of the possible are continually growing. In the longer term, expanding the repertoire of uncertainty visualisation formats, using robust methodology on a case-by-case basis, will improve the analysis and communication of uncertainty across a broad spectrum of decision-making contexts.

Finally, we must remember that **visualisation is not always the most appropriate tool.** Sometimes words or numbers do a better job of conveying a particular type of uncertainty, to a particular audience, in a particular context, for particular purposes. At the same time, we should be conscious that there is almost always some visual dimension involved in how we analyse and share information, how we design policies and processes, and how we imagine and plan for the future. Even when visualising uncertainty is not the main focus, being aware of the visual dimensions of a decision may be helpful in understanding and managing its uncertainties.

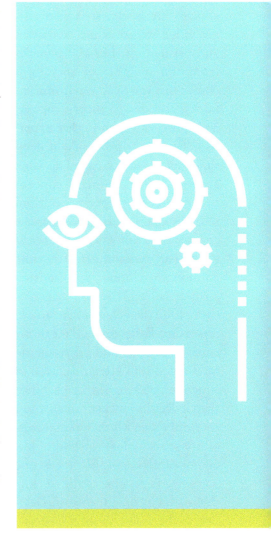

2 INTRODUCTION
THE INFLUENCE OF VISUALISATION

Decision theory is the study of how choices are made. It is primarily grounded in economics, statistics, and psychology, but also benefits from the insights of sociology, computer science, environmental science, design theory, the arts and humanities, and other expertise. The prescriptive side of decision theory—often called decision analysis—provides many conceptual resources to specify decisions formally and to create recommended courses of action. What counts as a 'rational' decision may vary from context to context, but decision analysis can help decision-makers to better fulfil their chosen standards of rationality. For example, if the future circumstances around a decision can be divided into scenarios, each of which can be assigned a probability, then a decision-maker can choose a course of action which is robust across all those scenarios—instead of just optimistically hoping for the best case scenario, or pessimistically steeling themselves against the worst. Decision analysis also allows us to anatomise a problem into its more tractable constituents, separating the technical from the value aspects. Then, it helps us to identify the role played by 'preferences' that must be generated by stakeholders through political processes, under conditions of more or less ethical legitimacy (or inferred by other means, e.g. in the case of future generations or non-human stakeholders).

Classifying, quantifying, and reasoning about uncertainty is central to decision analysis. So too is communicating about uncertainty. It has long been clear that not just what, but how we choose to communicate has considerable influence on the interpretations and actions that follow. Uncertainty can be communicated in different formats, including verbal descriptors ("high confidence"), numerical ranges, statistical graphics (e.g. probability density functions), pictograms, infographics, or various combinations. How uncertainty is communicated can be fundamental to the effective transfer of information between individuals, agencies, and organisations, and is thus crucial to decision-making. Visualisation may therefore play an especially significant role in analysis and decision-making involving multiple stakeholders. But even when only one person is responsible for every aspect of analysis and decision-making, visualisation may still play a significant role in how uncertainty is understood—i.e. in how the decision-maker 'communicates uncertainty to themselves.'

In many domains, there are already tried-and-true methods for classifying, quantifying, and propagating uncertainty in statistically robust ways. However, presenting uncertainty to decision-makers requires careful design and testing on a case-by-case basis. Visualisation can potentially influence decision-making in many ways, including:

- Whether or not a decision is made

- Which stakeholders input into the decision

- How evidence is prioritised

- Whether additional evidence is sought

- What forms of reasoning and analysis are used

- Whether biases are active and the extent of their influence

- How goals are formulated and success evaluated

- How much workload decision-making causes

- How long a decision takes

- Correctness of a decision

- Confidence in a decision

- Kinds of errors made

- How decision outcomes are interpreted

To support a decision, multidimensional uncertainty requires a representation that humans (acknowledging the variation in our capabilities) find natural to understand and operate. Decision-makers often use heuristics to assess uncertainty, and

Frequently, the challenge is to present visualisations in ways which allow the user intuitively to perceive the uncertainty probabilistically and arrive at effective decisions.

such assessments are often incompatible with probability theory, i.e. 'rational' treatment of uncertainty. Frequently, the challenge is to present visualisations in ways which allow the user intuitively to perceive the uncertainty probabilistically and arrive at effective decisions.

Visualisation can be used to counteract known biases, such as anecdotal evidence bias (Fagerlin et al., 2005), side effect aversion (Waters et al., 2006, Waters et al., 2007), and risk aversion (Schirillo & Stone, 2005). The effects of visualising uncertainty are frequently positive—but not always. Riveiro et al. (2014) write that 'even if the effects of visualizing uncertainty and its influence on reasoning are not fully understood, it has been shown that the graphical display of uncertainty has positive effects on performance.' Kinkeldey et al. (2015) observe: 'Overall, based on studies reviewed, uncertainty visualization has tended to result in a positive effect on decision accuracy. The evidence is less clear for decision speed, but it could be observed that usually, uncertainty visualization does not slow down decision-making and in one case it even decreased the decision time.'

Negative effects of visualisation reported in the literature point to interactions with cognitive biases and human psychology more generally, resulting in delays in decision-making when extra uncertainty-related information needs to be processed; irrational attitudes to risk such as focusing on the worst-case scenarios, or focusing on the mean rather than variance; confusion between risk and uncertainty; etc. Several studies explore how visualisation affects cognitive strategies for dealing with uncertainty, and find that visualisations might lead to less effort being expended by the decision-maker on acquiring and processing information relevant to uncertainty. This can be an undesirable consequence: Riveiro et al. (2014) report a study where visualisation of uncertainty in a military scenario led to significantly fewer attempts to identify a target, as well as higher threat values assigned to uncertain targets. This is sometimes explained by visualisations triggering an 'availability heuristic,' which puts undue emphasis on events that are readily imagined or easy to recall—such as the worst-case scenario—in disproportion to the chance of occurrence.

Training and experience of decision-makers plays a role in determining the impact of a particular visualisation. Kinkeldey et al. (2015) write that 'in order to use data and related uncertainty effectively, users must know how to interpret data together with related uncertainty.' Boone et al. (2018) conducted experiments on whether additional training on how to interpret specific graphical conventions for uncertainty visualisations could reduce flaws in decision-making. Using visualisation to express a hurricane's current location and projected path, together with uncertainty, they found that training can reduce misconceptions. However, one experiment also revealed an unexpected side-effect: reduced incidence of misinterpretation was correlated with lowered risk perception (decreased estimates of hurricane damage) relative to the group that did not receive training. Such evidence from past experiments reinforces the imperative to test new visualisation formats whenever possible.

Key ideas

decision theory	*stakeholders' preferences*
decision analysis	*multi-dimensional uncertainty*
robustness	*reduction of misconceptions through training*

3 THE CHALLENGES
THE STATE OF VISUALISATION RESEARCH

While there is solid evidence that visualisation influences decision-making, this influence is not uniformly positive, and theoretical models do not provide sufficient basis to predict case-specific impacts. Reviewing the state of knowledge in 2005, MacEachren et al. concluded that 'we do not have a comprehensive understanding of the parameters that influence successful uncertainty visualization, nor is it easy to determine how close we are to achieving such an understanding.' In a follow-up to that review in 2016, Riveiro asserted that this conclusion is still valid. This is not a reflection of the paucity of studies, as visualisation is a burgeoning research topic across diverse fields.

Indeed, a major difficulty in developing a broadly applicable theory is the sheer variety of methodological approaches and theoretical frameworks. Other common problems in the literature include small sample sizes for audience testing (statistical significance); inappropriate subjects (students rather than relevant decision-makers); lack of reproducibility; small effect sizes when comparing different visualisation approaches; biased self-perception (with little correspondence between self-reported impact and actual impact of visualisations); and transferability issues, confounded by the difficulty in controlling for differences in individual interpretation. Transferability is an especially salient problem since it implies that studies done with one group of people may not be applicable for another, or that the results are reproducible in general. Even on an individual level, responses to a visualisation may vary with factors such as time or stress-inducing constraints; the same visualisation may therefore influence decision-making in one way under a particular set of circumstances and in a contrary way under another.

When it comes to testing visualisation formats, there are broadly two types of investigations: *objective* (which rely on measured outcomes such as decision speed and decision accuracy) and *subjective* (which rely on self-reporting). Several studies offer evidence that self-reporting is unreliable, i.e. the user of a visualisation is not necessarily a good judge of how well or poorly the visualisation has supported their decision-making. This finding goes against the grain of a typical designer-client relationship, in which the designer has done a good job if the client is satisfied. In fact, client satisfaction may have little to do with the efficacy of the product: decisions can be improved by visualisations the client does not favour, or can be impaired by the visualisations the client happens to prefer. As often happens with good design, the benefits may not be noticeable to users. Several studies have reported that the users of visualisations may become better at decision-making without realising it. Kinkeldey et al. (2015) mention one study that revealed a striking lack of correlation between independently measured performance and self-reported confidence in making decisions: 'decision accuracy was significantly higher with uncertainty depicted, meaning that user performance and confidence did

Known problems

small sample sizes	*small effect sizes*
inappropriate subjects	*biased self-perception*
lack of reproducability	*transferability*

We do not have a comprehensive understanding of the parameters that influence successful uncertainty visualization, nor is it easy to determine how close we are to achieving such an understanding. (MacEachren et al., 2005)

not correspond.' The evidence pointing to users' inability to assess their own performance is broader than just research on visualisation, as Hullman et al. (2008) point out: 'evidence from other disciplines suggests that people are not very good at making accurate judgments about their own ability to make judgments under uncertainty.'

Even the level of expertise interacts with visualisation in ways which are difficult to anticipate. Greater expertise has been associated both positively and negatively with the effectiveness of visualisation. For example, in some studies a higher level of expertise enabled decision-makers to make better use of visualisation, arriving at decisions faster and with greater accuracy. In others, the level of expertise was associated with undesirable effects of visualisations; experiments 'showed that participants with a high level of experience had the strongest bias towards selecting areas of low uncertainty' (Kinkeldey et al., 2015).

Currently, much of the research on uncertainty visualisation relates to spatial reasoning. This research addresses the potential of visualisation to improve probabilistic spatial reasoning which, like all probabilistic reasoning, is plagued by cognitive biases. Some researchers, like Pugh et al. (2018), are optimistic: 'Visualizations have the potential to influence how people make spatial predictions in the presence of uncertainty. Properly designed and implemented visualizations may help mitigate the cognitive biases related to such prediction.'

Of course, creating visualisations that communicate uncertainty well is not necessarily a guarantee for effective decision support; even when understood correctly, uncertainty may be eschewed by decision-makers. Uncertainty is not something people are comfortable with no matter how well it is communicated. Resistance towards incorporating uncertainty into decision-making is widely reported in the literature. For example, Riveiro et al. (2014) report that 'participants' greater uncertainty awareness was associated with lower confidence.' Lower confidence, however, may be a desired outcome of visualisation in contexts where overconfidence is a known problem.

Kinkeldey et al. (2015) recommend that decision-makers are supported with more than just well-crafted visualisations: 'decision makers needed additional information for interpreting and coping with uncertainty (e.g. when a high degree of uncertainty is a problem and when not).' They point out that 'decision-makers often have little time to explore uncertainty in the data.' In order to ensure that decision-makers prioritise uncertainty information appropriately, the development of visualisation formats must be placed in a much wider context. This wider context includes the graphic literacy of decision-makers, the available ensemble of decision support tools, and the surrounding technical, social, and institutional infrastructures.

In short, what is lacking currently is the ability to predict how specific visualisations will impact particular decision-making processes. Uncertainty visualisation can be rewarding, but it is also a challenging and unpredictable territory. This is not to say that experience and existing research offer no guidance whatsoever, but to emphasise that any new visualisation needs to be tested and evaluated in its intended context and with the relevant audience, using robust methodology with an objective component (not just subjective appraisal). With these caveats in mind, the following sections present some basic approaches and practical examples of visualising uncertainty for decision support.

4 THE FRAMEWORK
DEVELOPING VISUAL SOLUTIONS

Before we begin discussing the foundational elements that can be called upon in the visualisation task, let's emphasise that selecting a visualisation method is not the first step. The process itself should begin with the identification of uncertainty, understanding of the various components that contribute to uncertainty, and discussing the aims of visualisation. We recommend considering the framework below or a close equivalent.

As emphasised in the previous section, self-reporting is an unreliable method for testing uncertainty visualisations. Decisions may be improved by visualisations the user does not favour, or may be impaired by the visualisations the user regards as helpful. This can create additional challenges within the design thinking process. On the one hand, as Deitrick and Wentz (2015) warn, visualisation research has often been 'normative in nature, reflecting what researchers think decision makers need to know about uncertainty,' instead of setting aside preconceptions and building empathy. On the

other hand, although designers and researchers must cultivate a deep understanding of decision-makers' lived experience, they must also work with decision-makers to explore how this experience may be misleading, once information from objective testing has been incorporated. The literature on bounded rationality, heuristics, and cognitive biases offers useful concepts in this regard (see section 08 "The User").

The following twelve-step guide demonstrates how design thinking can be implemented in the domain of uncertainty visualisation. It is largely based on A. Lapinsky's 'Uncertainty Visualization Development Strategy (UVDS).'

Step 1 is to classify the nature of the uncertainty. We explore possible approaches in the next section. Whatever the approach chosen, it will likely reveal that only a portion of uncertainty lends itself to visualisation. Deitrick and Wentz (2015) list common assumptions about the conditions under which uncertainty can be effectively visualised:

'First, it is assumed that uncertainty, or at least uncertainty of interest, is both knowable and identifiable. Similarly, to be visualized, uncertainty must be quantifiable, such as through statistical estimates, quantitative ranges, or qualitative statements (e.g. less or more uncertain). Moreover, evaluations define effectiveness as an ability to identify specific uncertainty values, which assumes that identifying specific uncertainty values is useful to decision-makers and that the values of interest can be quantified. Lastly, there is an assumption that the quantification of uncertainty is beneficial, applicable to the decision task, and usable by the decision maker, even if users do not currently work with uncertainty in that way.'

These pervasive assumptions mean that deep uncertainty, i.e. uncertainty that cannot be quantified given available resources, poses special challenges for visualisation.

Following Step 1, Steps 2 to 9 are the research phase, subdivided into 'Understand,' 'Decide,' and 'Determine.' Step 2 ensures that the data

Key ideas

i

self-reporting is unreliable

visualisation research shout set aside preconceptions and build empathy

literature on bounded rationality, heuristics, cognitive biases offers useful concepts

...although designers and researchers must cultivate a deep understanding of decision-makers' lived experience, they must also work with decision-makers to explore how this experience may be misleading

Figure 01.
12–Step Strategy for
Uncertainty Visualisation.
Based on the Uncertainty
Visualization Development
Strategy (UVDS) by Anna-
Liesa S. Lapinsky (2009).
Created by Jana Kleineberg.

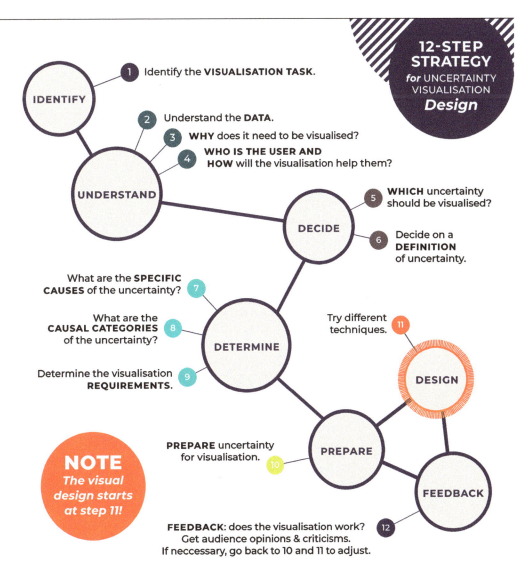

itself is understood, and takes into account such things as the origin and precision of the data, whether it has been modified (e.g. processed or aggregated), its format and the format's limitations, as well as other factors. Steps 3 and 4 look at the intended target audience: who will

use the uncertainty construct? How is it used? How will the visualisation help? It is important to determine the purpose for which presentation is required. Is it to explore the problem? To facilitate understanding between different stakeholders, roles, or forms of expertise? To analyse a situation

and make decisions? Some other purpose(s)? Step 5 helps decide on an information hierarchy, if multiple uncertainties are associated with the inputs. Step 6 formalises a definition of the uncertainty. Step 7 looks at the specific cause of uncertainty. Step 8 determines causal categories. Step

Decisions may be improved by visualisations the user does not favour, or may be impaired by the visualisations the user regards as helpful.

9 determines visualisation requirements, or the needs of the visualisation—e.g. what should the dominant features be, what level of understanding does the user have, and how will this influence the necessary level of detail? What tasks does the user need to perform, and what information is relevant to these tasks?

Step 10 leads into the actual design process by preparing the data for visualisation: sorting and organising measurements, converting them if necessary, converting uncertainty from collected data, and/or combining multiple uncertainties. Only in Step 11 does the creative part of the visualisation start. The principle of appropriate knowledge and the semantic principle provide guidance here. Different techniques can be used to create visualisation formats, encoding information in ways likely to support effectively attention and analysis. For example, a saliency algorithm (Padilla, Quinan, et al., 2017) can optionally be used to identify elements likely to attract viewers' attention. Once candidate visualisation formats have been developed, these are then tested in Step 12, ideally with the intended end-users themselves, employing evaluation methodologies that are transparent and reproducible. Steps 10 to 12 are connected, as they may be repeated multiple times to refine and improve on the visualisation.

Following general semantic and other design principles is no guarantee that visualisations will be correctly interpreted. Rather, a holistic approach to uncertainty visualisation includes principles of design, the testing of visualisation methods, the graphic literacy of end-users, as well as the overall decision-making context.

DESIGN THINKING

In graphic design the 'design thinking' approach has been widely adopted. It refers to the cognitive, strategic, and practical processes designers use to tackle complex problems. It is a flexible approach, focused on collaboration between designers and users, with an emphasis on bringing ideas to life based on how intended users think, feel, and behave.

The process is carried out in a non-linear fashion: the five stages are not always sequential and can often occur in parallel and/or iteratively. However, the design thinking model identifies and systematises the five stages one would expect to carry out in a design project.

Empathy: Understanding human needs; learning about the user who will interact with the design.

Definition: Framing and defining the problem; shaping a point of view based on user needs and insights.

Ideation: Creating ideas and creative solutions in ideation sessions, e.g. through brainstorming.

Prototyping: Adopting a hands-on approach by building (a) representation(s) to show to others.

Testing: Developing a solution to the problem and returning to the original user group for testing and feedback. Results are used to review the empathy stage, redefine problems, and refine the design.

DEEP UNCERTAINTY

Deep uncertainty may refer to uncertainty which we cannot quantify, which we cannot quantify given our available resources, or whose quantification is on balance undesirable.

So the distinction between uncertainty and deep uncertainty is not rigid, but rather is a function of the methods, resources, and choices we bring to bear on classifying and quantifying uncertainty. Identifying some uncertainties as deep uncertainties does not place them beyond quantification once and for all, and does not rule out the possibility that they may be reclassified at a later stage in the process.

Furthermore, adopting the designation of deep uncertainty does not mean that decision-makers are justified in excluding these uncertainties from their reasoning, or that developers of decision support tools can safely set them to one side.

Indeed, it is even theoretically possible to visually depict deep uncertainty: many artworks (e.g. the abstract impressionist works of Rothko) might be considered representations of deep uncertainties that are perceived intuitively. However, in line with the majority of research done to date, this primer focuses on uncertainties which can be quantified.

5 | TYPES OF UNCERTAINTIES
CATEGORISING UNCERTAINTY

WHY CATEGORISE?

Decision-making can be improved by understanding the uncertainty in the data being used. Categorising uncertainty is a preliminary step towards recognising and dealing with uncertainty in the decision-making process.

Uncertainty can come in many forms, and with many different qualities. Each type of uncertainty applies to different types of information, and may be quantified, and thus represented, in different ways. However, there is no one-to-one correlation of uncertainty types and visualisation techniques. As Chung and Wark (2016) confirm, often the same technique, e.g. colour coding, has been variously applied to depict distinct types of uncertainty, as well as subsets of combined/compounded uncertainties from different categories.

Uncertainty has been decomposed in a variety of ways by a multitude of theoretical treatments, and each classification scheme carries its own set of assumptions and motives. In particular, sometimes uncertainty is broken up into different types simply to help us spot where uncertainty lies, and to organise how we investigate and manage uncertainty, but on the understanding that these distinctions don't ultimately prevent integration into a single analysis and decision procedure. On other occasions, the purpose of a classification scheme is to draw attention to more fundamental differences between uncertainty types, which may be difficult or impossible to reconcile.

Key ideas

categorizing is key

uncertainty comes in many different forms & qualities

understanding is tailored to context

Uncertainty can come in many forms, and with many different qualities.

SOME PROMPTS

French et al. (2016) list the following types of uncertainty: **stochastic** uncertainties (i.e. physical randomness), **epistemological** uncertainties (lack of scientific knowledge), **endpoint** uncertainties (when the required endpoint is ill-defined), **judgemental** uncertainties (e.g. setting of parameter values in codes), **computational** uncertainties (i.e. inaccurate calculations), and **modelling errors** (i.e. however good the model is, it will not fit the real world perfectly, or if it seems to, it is likely to have little predictive power). There are further uncertainties that relate to **ambiguities** (ill-defined meaning) and partially formed **value judgements**; and then there are **social** and **ethical** uncertainties (e.g. how expert recommendations are formulated and implemented in society, what the ultimate ethical value of a decision and all its consequences will be). Some uncertainties may be deep uncertainties; that is, within the time and data available to support the decision process, there may be little chance of getting agreement on their evaluation or quantification.

Chung and Wark (2016) list these categories in their review of uncertainty visualisation literature:

- **Accuracy** *the difference between observation and reality*
- **Precision** *the quality of the estimate or measurement*
- **Completeness** *the extent to which information is comprehensive*
- **Consistency** *the extent to which information elements agree*
- **Lineage** *the pathway through which information has been passed*
- **Currency** *the time span from occurrence to information presentation*
- **Credibility** *the reliability of the information source*
- **Subjectivity** *the extent to which the observer influences the observation*
- **Interrelatedness** *the dependence on other information*
- **Experimental** *the width of a random distribution of observations*
- **Geometric** *the region within which a spatial observation lies*

The two categorisations presented illustrate the breadth of approaches to uncertainty, and suggest that the choice of a schema for understanding uncertainty might need to be tailored to the context.

Excerpt from "Decision Support Tools for Complex Decisions under Uncertainty," edited by Simon French from contributions from many in the AU4DM network:

6 CYNEFIN

Another approach to uncertainty called Cynefin — a Welsh word for habitat, and used here to describe the context for a decision — categorises our knowledge relative to a specific decision. Cynefin roughly divides decision contexts into four spaces (see figure 02). Note that placing a decision in one of these four spaces does not preclude certain aspects of that decision being associated with a different space. It may also occasionally be appropriate to situate a decision in a particular space for one set of purposes, and in a different space for another set of purposes. Acquiring more information and/or conducting analysis may also shift a decision from one space into another.

In the **Known Space**, also called *Simple*, or the *realm of Scientific Knowledge*, relationships between cause and effect are well understood, so we will know what will happen if we take a specific action. All systems and behaviours can be fully modelled. The consequences of any course of action can be predicted with near certainty. In such contexts, decision-making tends to take the form of recognising patterns and responding to them with well-rehearsed actions, i.e. recognition-primed decision-making. Such knowledge of cause and effect will have come from familiarity. We will regularly have experienced similar situations. That means we will not only have some certainty about what will happen as a result of any action, we will also have thought through our values as they apply in this context. Thus, there will be little ambiguity or value uncertainty in such contexts

In the **Knowable Space**, also called *Complicated*, or the *realm of Scientific Inquiry*, cause and effect relationships are generally understood, but for any specific decision further data is needed before the consequences of any action can be predicted with certainty. The decision-makers will face epistemological uncertainties and probably stochastic and analytical ones too. Decision analysis and support will include the fitting and use of models to forecast the potential outcomes of actions with appropriate levels of uncertainty. Moreover, although the decision-makers will have experienced such situations before they may be less sure of how their values apply and will need to reflect on these in making the final decision.

REPEATABILITY AND INCREASED FAMILIARITY

Simple/Known space
The realm of Scientific Knowledge, also called the 'known knowns.' Rules are in place, the situation is stable. Cause and effect relationships are understood; they are predictable and repeatable.

Complicated/ Knowable space
The realm of Scientific Inquiry, or domain of 'known unknowns.' Cause and effect relationships exist. They are not self-evident but can be determined with sufficient data.

'MESSY' DECISIONS, MANY UNCERTAINTIES ARE DEEP

Complex space
The realm of Social Systems, or domain of 'unknown unknowns.' Cause and effect are only obvious in hindsight and have unpredictable, emergent outcomes.

Chaotic space
No cause and effect relationships can be determined.

Cynefin: Welsh, without direct translation into English, but akin to a place to stand, usual abode, and habitat. It is pronounced /ˈkʌnivin/ KUN-iv-in.

In the **Complex Space**, also called the *realm of Social Systems*, decision-making faces many poorly understood, interacting causes and effects. Knowledge is at best qualitative: there are simply too many potential interactions to disentangle particular causes and effects. There are no precise quantitative models to predict system behaviours such as in the Known and Knowable spaces. Decision analysis is still possible, but its style will be broader, with less emphasis on details, and more focus on exploring judgement and issues, and on developing broad strategies that are flexible enough to accommodate changes as the situation evolves. Analysis may begin and, perhaps, end with much more informal qualitative models, sometimes known under the general heading of soft modelling or problem structuring methods. Decision-makers will also be less clear on their values and they will need to strive to avoid motherhood-and-apple-pie objectives, such as minimise cost, improve well-being, or maximise safety.

Contexts in the **Chaotic Space** involve events and behaviours beyond our current experience and there are no obvious candidates for cause and effect. Decision-making cannot be based upon analysis because there are no concepts of how to separate entities and predict their interactions. The situation is entirely novel to us. Decision-makers will need to take probing actions and see what happens, until they can make some sort of sense of the situation, gradually drawing the context back into one of the other spaces.

The **central blob** in figure 02 is sometimes called the Disordered Space. It simply refers to those contexts that we have not had time to categorise. The Disordered Space and the Chaotic Space are far from the same. Contexts in the former may well lie in the Known, Knowable, or Complex Spaces; we just need to recognise that they do. Those in the latter will be completely novel.

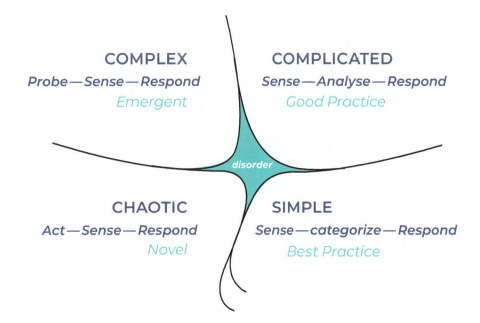

Figure 02. Uncertainty Cynefin

For more information on categories of uncertainty and Cynefin, please see the AU4DM Uncertainty catalogue "Decision Support Tools for Complex Decisions under Uncertainty" edited by Simon French from contributions from many in the AU4DM network.

7 | VISUALISING UNCERTAINTY
THE TOOLKIT

As noted, developing an uncertainty visualisation solution begins with discussing the needs of the user, and understanding the various components that contribute to uncertainty. Only once this has been done can assessment of appropriate visualisation techniques begin. Furthermore, since there is no general theory of uncertainty visualisation, whichever techniques we select will still need to be tested, potentially through multiple iterations. Testing should ideally be undertaken with the actual end-users of the uncertainty visualisation. Given well-attested problems with self-reporting, testing should also not rely solely on users' subjective experience. In summary, any toolbox sits within a larger process of preparation and testing. There are various credible approaches to managing this process; this catalogue proposes an approach described in Section 4.

CLASSIFYING VISUALISATION TECHNIQUES

There are a variety of classification schemes for visualisation techniques used in uncertainty visualisation. Deitrick (2012) distinguishes between implicit and explicit visualisation of uncertainty information. Uncertainty is implicitly visualised when encoded in the image in such a way that uncertainty cannot be separated out as a feature: that is, no design element represents uncertainty by itself without also signifying some other value. 'Explicit visualization refers to methods where uncertainty is extracted, modeled and quantified separately from the underlying information' (Deitrick, 2012). As an example of implicit visualisation, Deitrick offers a scenario in which a decision-maker is reviewing three graphics associated with three policy options. In each graphic, the vertical and horizontal axes represent two variables whose future state is not known, and the graph space is shaded to represent how successful the policy will be for any given combination of values. In the explicit version, by contrast, there is an underlying model to predict the outcome of each policy, and each visualisation encodes the model's uncertainty in the transparency/opacity dimension. Experiments suggest that visualising uncertainty implicitly versus explicitly can impact decision-making in divergent ways (Deitrick, 2012). In this catalogue, we generally focus on explicit visualisations.

Kinkeldey et al. (2014) review an array of studies in terms of how uncertainty is visualised, classifying approaches to visualisation according to three theoretical dichotomies. (i) Coincident/adjacent distinguishes information represented together with its uncertainty (coincident) from information and associated uncertainty that are visualised separately (adjacent). (ii) Intrinsic/extrinsic distinguishes visualisations achieved through manipulation of existing graphical elements (intrinsic), from

Key ideas

coincident/adjacent	*selective attention & visual salience*
intrinsic/extrinsic	*using colour to visualise uncertainty*
static/dynamic	*(introducing some examples)*

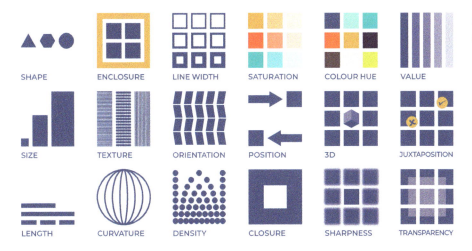

SHAPE ENCLOSURE LINE WIDTH SATURATION COLOUR HUE VALUE

SIZE TEXTURE ORIENTATION POSITION 3D JUXTAPOSITION

LENGTH CURVATURE DENSITY CLOSURE SHARPNESS TRANSPARENCY

Figure 03. Attributes of Graphics. Crated by Jana Kleineberg.

those derived through addition of elements such as grids, glyphs, and icons (extrinsic). (iii) Finally, static/ dynamic refers to the potential of a visualisation to change through time, as with animation or interactive techniques. Pragmatic approaches by Meredith et al. (2008) and Matthews et al. (2008) are motivated by the question 'what can be done to visu- alise uncertainty?' In their review of the approaches to the visualisation of uncertainty, Meredith et al. (2008) mention the following:

- adding glyphs*
- adding icons**
- adding geometry
- modifying geometry
- modifying attributes
- animation
- sonification
- fluid flow
- surface interpolants
- volumetric rendering
- differences in tree structures

glyphs: In typography, a glyph is an elemental symbol within an agreed set of symbols, i.e. an individual mark of a typeface, such as a letter, a punc- tuation mark, an alternate for a letter. A glyph is usually a mark that repre- sents something else. For example, the @ sign is a glyph that commonly represents the word 'at'. Thus, e.g. in flow fields, one would speak of glyphs (usually pointing arrows) that show a trend or a direction, as there cannot be a direct pictorial representation of "flow".

**icons:* An icon is a more direct representation of something else; a pictogram or ideogram that shows a simplified, comprehensible symbol of the function or thing it represents. Icons are often recognisable depic- tions of familiar objects, such as fish, cars, or trees. The sections below will provide examples of glyphs and icons.

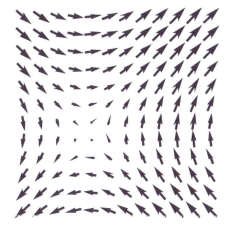

Figure 04. Vector field, or flow field. Source: Wikipedia.

Matthews et al. (2008) offer four nested categories of techniques: alter- ation (1), addition (2), animation (3), and interaction (4):

Free graphical variables (e.g. colour, size, position, focus, clarity, fuzzi- ness, saturation, transparency and edge crispness) can be used to alter aspects of the visualizations to communicate uncertainty.

1. Additional static objects (e.g. labels, images, or glyphs) can be added to the visualizations to communicate uncertainty.

2. Animation can be incorporated into the visualizations, where uncertainty is mapped to animation parameters (e.g. speed, duration, motion blur, range or extent of motion).

3. Uncertainty can be discovered by mouse interaction (e.g. mouse-over).'

There are only so many graphical attributes that can be manipulated (figure 03). Various combinations of graphical parameters have been explored in the context of visual- ising uncertainty. Individual stud- ies and reviews of existing research offer valuable insight into their usefulness.

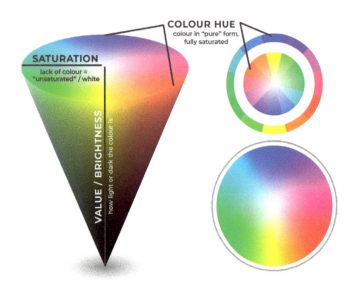
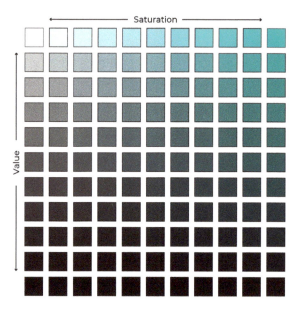

Figure 05. Hue, Saturation, Value. Created by Jana Kleineberg

SELECTIVE ATTENTION AND VISUAL SALIENCE

Certain visual attributes can have a significant impact on our so-called 'preattentive' perception. For example, certain stimuli tend to 'jump out' at us, regardless of our top-down preferences, strategies, and goals in processing visual information. Influences on preattentive perception:

- **Colour** hue, saturation, and value

- **Size** the surface area of an element

- **Position** where an element sits within a visualisation's overall space and/or various subspaces

- **Predictability** whether the element is in its expected position

- **Set size** the total number of elements in a visualisation

- **Emotional connotations** e.g. smiling or angry faces, 'cute' imagery of animals or babies

- **Contrast** more broadly, how an element compares to other elements in the visualisation as regards these attributes and others

- **Visual salience** more broadly still, the conspicuousness of an element relative to its environment, influenced by all of the above as well as other factors, e.g. motion, sharpness of edges, orientation

Manipulation of these features may impact the allocation of attention, e.g. the likelihood that the decision-maker notices an element at all, and their likelihood of fixating on it. The term visual salience refers to the conspicuousness of a visual element relative to the visual surroundings in which it appears (Itti and Koch, 2000). A salience model takes as input any visual scene and produces a topographical map of the most conspicuous locations, i.e. those locations that are brighter, have sharper edges, or different colours than their surroundings. The saliency map (Koch & Ullman, 1985) usually considers three channels: colour, intensity, and orientation—drawn from a variety of different spatial scales. The map itself represents the visually most important regions in the image. Under normal viewing conditions, there is believed to be a positive association between the location of fixation and salience of the stimulus at those locations. In other words, salience exerts a small but significant effect on fixation likelihood, so that decision-makers are more likely to fixate on objects with a greater level of salience (Milosavljevic, Navalpakkam, Koch, and Rangel, 2012). for instance, Milosavljevic et al. (2012) demonstrated that, during rapid decision-making tasks, visual saliency influences choices more than personal preferences (Kahneman et al., 1982). Such salience bias increases with cognitive load and is particularly strong when individuals do not have strong preferences related to different options. Salience has also been shown to influence the fixation order, with more salient objects being fixated on earlier (Peschel and Orquin, 2013).

ANY COLOUR CAN BE DEFINED ACCORDING TO ITS POSITION IN THREE DIMENSIONS: HUE, VALUE, AND SATURATION

We should take care to note some terminological inconsistency, especially around descriptions of colour. for example, other terms for saturation include intensity and purity. Other terms for value include brightness, lightness, and luminosity. All these terms risk being confused with transparency, which is not really a perceptual dimension of the colour itself, but the degree to which the colour(s) underneath are allowed to show through. Sometimes the terms colour and hue are used interchangeably. for example, in a review of visualisation of uncertainty by Aerts et al. (2003) we read: 'Bertin (1983) describes an extended set of visual variables to portray information, such as position, size, value, texture, color, orientation, and shape. Among these variables, "the strongest acuity in human visual discriminatory power relates to varying size, value and color".' Colour in this context may refer to hue, or to a combination of hue and saturation. As well as terminological inconsistency, there is not infrequently some conceptual confusion associated with these dimensions of colour.

USING COLOUR TO VISUALISE UNCERTAINTY

Any colour can be defined according to its position in three dimensions: hue, value, and saturation. Figure 05 illustrates the relationship between these three perceptual dimensions. Studies such as Seipel and Lim (2017) explore the use of hue, saturation, and value in communicating uncertainties. However, Kinkeldey et al. (2014) report that 'from current knowledge, colour saturation cannot be recommended to represent uncertainty. Instead, colour hue and value as well as transparency are better alternatives.'

There is another mention of saturation as an unsuitable graphical parameter for the visualisation of uncertainty in Cheong's 2016 review, although the same studies are considered in both reviews and so there is a risk of double counting and drawing stronger conclusions than the evidence supports. Cheong (2016) further notes inconsistencies across various inquiries into the use of hue and value, with some experiments confirming effectiveness while others dispute it; these apparent conflicts are likely due to studies not being directly comparable, e.g. in terms of experimental subjects. Where value is effective, the literature suggests that

people tend to associate darker values with more certainty, and lighter values with less (e.g. MacEachren, 1992).

Hue is determined by the dominant wavelength, and is the term that describes the dimension of colour we first experience when we look at a colour ("yellow," "blue," etc.). When we speak of hue, we are generally referring to the colour in its "pure" and fully saturated form. Saturation refers to how pale or strong the colour is. To simplify just a little, it can be said that the more white you add, the less saturated the colour becomes. Value refers to how light or dark a colour is. Adding grey or black will

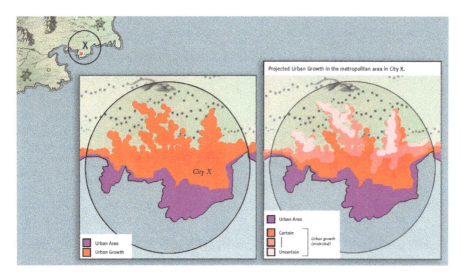

Figure 06 (left). Illustrating value and uncertainty in separate representations, arranged side-by-side.

Figure 07 (bottom). Using hue to directly illustrate uncertainty.

Figure 08 (page 21). Comparing changing different attributes to communicate uncertainty. Based on Cheong et al. (2016).

All created by Jana Kleineberg

change the value: a low value is dark grey or black, and a high value is light grey or white.

A distinction is also made between pigment primaries (e.g. print) and light primaries (e.g. pixels). Pigment primaries use subtractive colour mixing: dyes, inks, paints, pigments absorb some wavelengths of light and not others. Typical 'backgrounds,' such as fabric fibres, paint base, and paper without pigments, are usually made of particles that scatter all colours in all directions, meaning they look white. When a pigment or ink is added, specific wavelengths are absorbed (subtracted) from white light, so light of another colour reaches the eye (the colour we see). The primary colours of this colour model are cyan, magenta, and yellow (CMY); combining all three pigment primary colours yields black. By contrast, light on a monitor display, projector, etc. uses additive colouring: mixing together light of two or more different colours. The primary colours of this colour model are usually red, green, and blue (RGB). All three primary colours together yields white.

VISUALISING UNCERTAINTY: SOME EXAMPLES

Figures 06 to 11 give some examples to illustrate these techniques. Figure 07 demonstrates the use of hue to convey uncertainty about spatial values. This fictitious example shows the projected territorial range of an invasive species. The key or legend, an essential feature of graphical displays, explains how two hues convey the probabilities that a species will spread to respective areas.

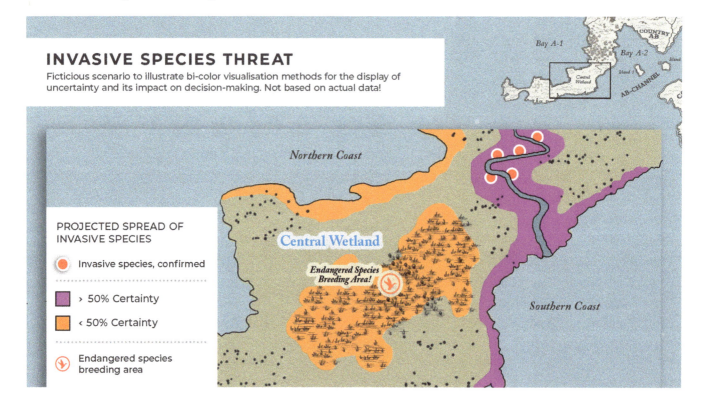

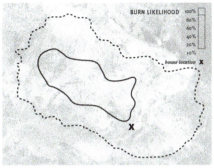
a. Boundary

b. Colour hue

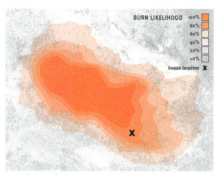
c. Colour value

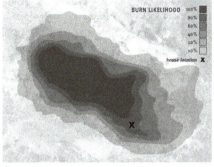
d. Transparency

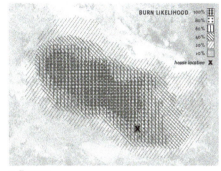
e. Texture

Your house is located in the
>80 to 100%
burn likelihood zone.

f. Text

Some studies have suggested using side-by-side representations of the variable and uncertainty relating to the variable, e.g. Deitrick and Wentz (2015). For illustration purposes, consider Figure 06, where the left panel depicts projected future size of a metropolitan area of a fictional city and the right panel shows uncertainty in these projections.

An alternative to a side-by-side display is an interactive display, enabling users to switch between representations of a variable and uncertainty. Various studies, involving high stakes, high uncertainty, and time pressure, have shown that in simulations the ability to switch between alternative representations of uncertainty is helpful. Finger and Bisantz (2000) explored communicating uncertainty in radar contacts by degrading or blurring the icons used to represent them. Bisantz et al. (2011) expanded this research, as Riveiro et al. (2014) summarise: 'several display methods were used in a missile defense game: icons represented the most likely object classification (with solid icons), the most

likely object classification (with icons whose transparency represented the level of uncertainty), the probability that the icon was a missile (with transparency) and, in a fourth condition, participants could choose among the representations. Task performance was highest when participants could toggle the displays, with little effect of numeric annotations. As such, the authors once more support the use of graphical uncertainty representations, even when numerical presentations of probability are present.'

Research suggests that representations involving hue (b), value (c) and transparency (d) worked best (Figure 07). In addition to transparency, value, and hue, graphical attributes that were found useful in representing uncertainty include resolution, fuzziness, and blurring (Kinkeldey et al., 2014).

Riveiro et al. (2014) concur, citing a couple of visualisation of uncertainty evaluations where 'fuzziness and location seem to work particularly well, and both size and transparency are potentially usable.'

These approaches make use of metaphors. Kinkeldey et al. 2014 write: 'The contention is that fog and blur are metaphors for lack of clarity or focus (as in a camera) and thus directly signify uncertainty. These metaphors have been suggested to have the potential to enable a better understanding of uncertainty (Gershon, 1998) and we make the assumption that the use of metaphors can lead to more intuitive approaches.' Metaphors can be both useful and misleading: for example, particular colour hues carry connotations which might interfere with intended signification. For example, discussing climate change modelling visualisations, Harold et al. 2017 warn against the use of blue that may be misinterpreted as representing water (Figure 11).

Metaphors are not universal, and associations might differ depending on the culture and experiences of users. Further, as Kinkeldey et al. (2015) show, the choice of colour hue can have an impact on the perception of risk, and hence on decision-making

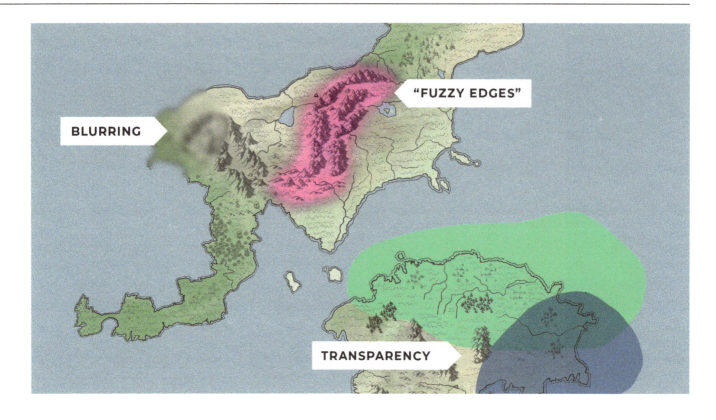

under uncertainty, such as the decision about whether or not to follow an evacuation order. To understand the impact of colour hue we must understand the emotional significance for users, which might be different for different users based on individual and group-linked factors. Further, the impact of the same colour hue on a person's decision-making might depend on the circumstances, for instance triggering a different set of heuristics when the person is under pressure. The choice of colour hue on a map could translate into a number of

lives lost if design influences people's decision not to follow an evacuation order, where some other hue would have better conveyed an appropriate level of urgency. In situations where an audience for the visualisation is diverse but it is impossible to tailor visualisations to distinct groups (as with hurricane warnings), it may not be trivial to make choices regarding visualisation formats, as trade-offs between the overall efficacy and group specific impacts might need to be considered.

Figure 09 (above).
Using blurring, fuzziness, and transparency to communicate uncertainty.

Figure 10 (bottom).
Using pixelation to represent uncertainty.

Figure 11 (page 23).
Colour and metaphor. Based on Harold et al. 2017. The colour blue is traditionally used to represent water; thus data might not be read correctly.

All created by Jana Kleineberg.

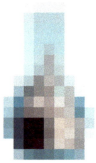

EFFECTIVE GRAPHIC DESIGN PRINCIPLES
Boone et al. (2018) list the following principles of effective graphic design:

'Effective graphic design takes account of

- the **specific task** at hand (Hegarty, 2011)

- **expressiveness** of the display (Kosslyn, 2006)

- data-ink **ratio** (Tufte, 2001)

- issues of **perception** (Kosslyn, 2006; Tversky, Morrison, & Betrancourt, 2002; Wickens & Hollands, 2000)

- **pragmatics** of the display, including making the most relevant information salient (Bertin, 1983; Dent, 1999; Kosslyn, 2006)

It also takes account of semantics:

- **compatibility** between the form of the graphic and its meaning (Bertin, 1983; Kosslyn, 2006; Zhang, 1996)

- **usability** of the display, such as including appropriate knowledge (Kosslyn, 2006)'

Ignoring these principles, or failing to implement them effectively, may lead to misunderstandings, or other forms of suboptimal decision support.

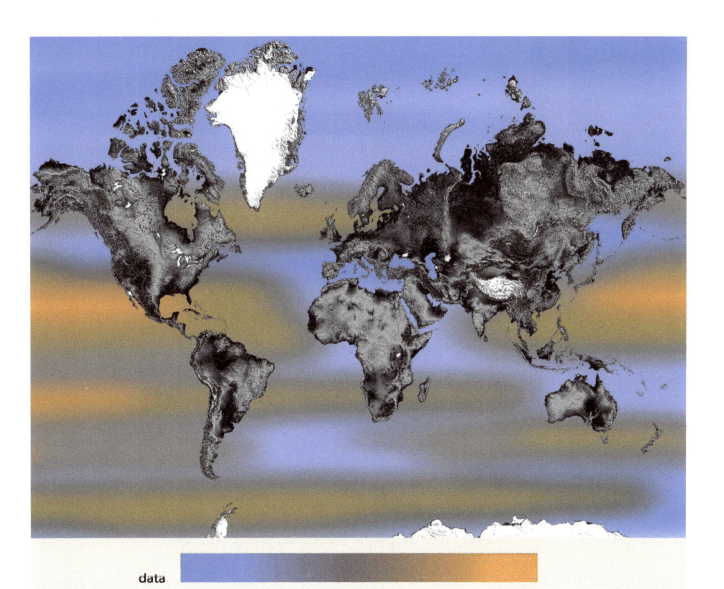

data

8 | THE USER

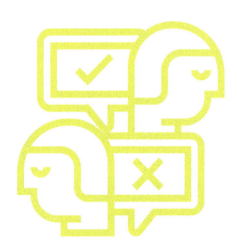

In order to effectively visualise uncertainty, it is necessary to understand how mistakes can occur in reception. As emphasised throughout this catalogue, such understandings should be developed through empirical testing with users. However, two design principles do offer some guidance: the principle of appropriate knowledge (which relates to the user's familiarity with conventions) and the semantic principle (which relates to 'natural' mappings between visualisations and visualised information).

Furthermore, it is useful to appreciate that a visualisation format may have several different kinds of user, as well as stakeholders who rely on it in a more indirect fashion. In particular, it is useful to be aware of how the persuasive power of visualisation can play out in distributed decision-making settings, and how it may interact with asymmetries of information, expertise, experience, authority, and accountability among analysts, decision-makers, and other stakeholders.

Finally, the process of developing visualisation formats for decision support can be informed by a range of different models of cognition and decision-making, including an understanding of the decision-maker as a boundedly rational agent, an understanding of common heuristics and biases related to perceiving and reasoning about uncertainty, and an understanding of the counterintuitive quirks of visual perception (i.e. those that underlie optical illusions). In this section we briefly touch on these topics.

Keywords

principle of appropriate knowledge	clustering illusion	framing effect
semantic principle	priming	confirmation bias
visualisation & persuasion	anchoring effects	set size effect
heuristics & cognitive biases	Weber's law	surface size effect
confounding mean and variance	Ostrich effect & risk compensation bias	position effects & predictable locations
colour contrast phenomena	availability bias	emotional stimuli
		inter-individual variability

The user of a visualisation must have knowledge of the conventions necessary to extract its relevant information.

THE PRINCIPLE OF APPROPRIATE KNOWLEDGE AND THE SEMANTIC PRINCIPLE

Heuristics are short-cut procedures or rules of thumb that may generate good-enough results under certain conditions but are also implicated in generating biased decisions. Cognitive biases are systematic errors in one's thinking relative to either social norms for reasoning and/or formal logic. Without testing, it is not possible to know how a particular visualisation format will interact with a user's heuristics and biases. However, some principles of good design practice exist, and are applicable to a wide range of visualisation formats and their users. Two such principles, which are related to each other, are the principle of appropriate knowledge and the semantic principle.

The principle of appropriate knowledge simply states that the user of a visualisation must have knowledge of the conventions necessary to extract its relevant information. The conventions of the display are often encoded in a legend expressing the correspondence between visual variables and their meaning. There may also be additional instructions, cautions, or recommendations to help with interpretation. The principle of appropriate knowledge also invites us to think about the risk that users will interpret a visualisation based on inappropriate conventions. If the user does apply a different convention to the one intended, will this produce dissonance, causing the user to feel that something is wrong? Or will the visualisation apparently accommodate the incorrect convention, allowing the user to incorrectly interpret the visualisation indefinitely? Furthermore, the principle of appropriate knowledge invites us to think carefully about the conventions we rely on in encoding and interpreting information visually, since practices that feel obvious or inevitable may actually rely on norms that have been learned at some point, and that are not necessarily shared by all users.

In practice, it may not be possible to know the user's familiarity with various conventions. This is where the semantic principle comes in. In the context of creating visualisations of uncertainty, Boone et al. (2018) describe 'the semantic principle of natural mappings between variables in the graphic and what they represent.' As examples, they offer 'classic metaphors such as "larger is more" and "up is good" (Tversky, 2011).' According to the semantic principle, visual attributes should be mapped to underlying data in 'common sense' ways, so that most users would correctly guess how to interpret it, even without knowing the conventions used. 'An example of a match [between visualisation and underlying data] is using the length of a line to denote length of time. An example of a mismatch would be using higher values on a graph to show negative numbers' (Boone et al., 2018). Another example of a match would be data values that are related to one another being physically proximate. Padilla et al. (2018) recommend that we '[a]im to create visualizations that most closely align with a viewer's mental schema and task demands' and '[w]ork to reduce the number of transformations required in the decision-making process.'

The semantic principle raises some interesting theoretical questions (see sidebar). However, practically speaking, the chief drawback of the semantic principle is that it does not always provide a reliable or sufficiently detailed guide to support the visualisation of complex information such as uncertainty information. It is therefore usually

Connecting the First Mile

Investigating Best Practice for ICTs and Information Sharing for Development

Surmaya Talyarkhan
David J. Grimshaw
Lucky Lowe

Practical
ACTION
PUBLISHING

Practical Action Publishing Ltd
25 Albert Street, Rugby, CV21 2SD, Warwickshire, UK
www.practicalactionpublishing.com

© Intermediate Technology Publications 2005

First published 2005

ISBN 10: 1 85339 612 5
ISBN 13 Paperback: 9781853396120
ISBN Library Ebook: 9781780441498
Book DOI: https://doi.org/10.3362/9781780441498

A catalogue record for this book is available from the British Library.

The authors, contributors and/or editors have asserted their rights under the Copyright
Designs and Patents Act 1988 to be identified as authors of their respective contributions.

Since 1974, Practical Action Publishing has published and disseminated books and
information in support of international development work throughout the world.
Practical Action Publishing is a trading name of Practical Action Publishing Ltd
(Company Reg. No. 01159018), the wholly owned publishing company of Practical
Action. Practical Action Publishing trades only in support of its parent charity objectives
and any profits are covenanted back to Practical Action (Charity Reg. No. 247257, Group
VAT Registration No. 880 9924 76).

Reasonable efforts have been made to publish reliable data and information, but the
author and publisher cannot assume responsibility for the validity of all materials or for
the consequences of their use.

Typeset by J&L Composition, Filey, North Yorkshire

The manufacturer's authorised representative in the EU for product safety is
Lightning Source France, 1 Av. Johannes Gutenberg, 78310 Maurepas, France.
compliance@lightningsource.fr

Contents

Acknowledgements

This paper summarizes the findings from a two-year research project conducted by ITDG and Cranfield University into the use of information communication technologies (ICTs) for development. This project was managed as a Knowledge Transfer Partnership, funded in part by the Department of Trade and Industry.

The authors would like to thank staff at ITDG for providing materials and sharing their experiences. In particular, the staff of the SIRU project in Peru, who facilitated the fieldwork and were open about the lessons they had learned. The authors also benefited from valuable input from colleagues at Cranfield School of Management. Many thanks also to Pat Norrish, who kindly peer-reviewed the final draft.

Surmaya Talyarkhan is the Knowledge Sharing Adviser at ITDG, specializing in the field of information communication technologies for development and knowledge sharing.

David J. Grimshaw is the International Team Leader of the 'Responding to New Technologies Programme' at ITDG, and Visiting Fellow at Cranfield School of Management. His areas of expertise are: in information systems strategy, usability, knowledge management, ICTs for development, and geographical information systems.

Lucky Lowe is a Professional Manager with 20 years experience working in the UK construction and international human settlements sectors. More recently she worked on knowledge sharing issues as ITDG's Knowledge and Information Services Unit Manager.

Acronyms

CARE	Cooperativa Americana de Remesas a Europa
CEDEPAS	Centro Ecuménico de Promoción y Acción Social
CPI	Centro de Procesamiento de Información (Information Processing Centre)
DIA	Dirección de Información Agraria
DOT-Force	Digital Opportunities Task Force
DVD	Digital Video Disk
GIS	Geographical Information Systems
ICTs	Information Communication Technologies
IK	Indigenous Knowledge
INC	Instituto Nacional de Cultura (National Institute of Culture, Peru)
IPR	Intellectual Property Rights
ITDG	Intermediate Technology Development Group
OECD	Organisation for Economic Co-operation and Development
PDA	Personal Digital Assistant
PRA	Participatory Rural Appraisal
PRODELICA	Proyecto de Desarrollo Integral La Libertad – Cajamarca
PRONOMACH	Proyecto Nacional de Manejo de Cuencas Hidrográficas y Conservación de Suelos, Hydro project in Peru
RRA	Rapid Rural Appraisal
SIRA	Sistema de Información Rural de Arequipa
SIRU	Sistema de Información Rural Urbano (Rural-Urban Information System)
SNV	Dutch Development Agency
UN	United Nations
WiFi	Wireless Fidelity
WSIS	World Summit on the Information Society

Executive summary

Across the globe, development agencies are piloting projects to improve access to information in developing countries, many of which exploit the potential of information communication technologies (ICTs). Projects face the challenge of sharing information with people who have little experience of ICTs, low levels of literacy, little time or money, and highly contextualized knowledge and language requirements.

This paper characterizes this as the challenge of *'connecting the first mile'* and aims to answer the question "What is best practice in connecting the first mile?" through an analysis of the literature and a case study based on practical experience of an ICT for development project in Cajamarca, Peru and to offer recommendations for practitioners and suggestions for further research.

The role of information and knowledge in development is contentious. Whilst it is clear that information is central to development, practitioners struggle to define a causal link between information and development outcomes. In Section 2, we discuss the role of information and of knowledge in development and conclude that information sharing in itself does not necessarily lead to development outcomes, unless the processes are in place to support the transformation of information into knowledge.

ICTs can potentially play a valuable role in sharing information rapidly and effectively. Developments in the ICTs industry such as convergence and reducing costs lead to hopes that technology leapfrogging will help to bridge the 'digital divide'. In practice, ICT for development projects face criticism for being top-down or neglecting the local context or viewing development through a technological lens. Practitioners have difficulty identifying how a project has contributed to development goals, how to monitor and evaluate that contribution, how to ensure the project is sustainable, and how to 'scale up' or replicate a successful project in a different context. Section 3 discusses these issues in detail.

Evidence from existing projects suggests that that the success of many projects is 'situated success' in the sense that the project has worked due to a particular combination of local factors such as a strong champion or good timing. But there are also processes that most authors highlight as contributing to the success of projects in all contexts. Section 4 identifies the key factors that have contributed to the success of projects and the activities which constitute best practice. These are summarized below:

1. Start from the people, not the technology
   - Starting from community priorities
   - Understanding local power dynamics
   - Minimizing social exclusion
   - Strengthening social capital
2. Blend ICTs with traditional information systems
   - Building on existing information systems
   - Choosing appropriate technologies
   - Adapting to local infrastructure

3. Share information that users can appropriate
   - Researching information needs
   - Developing appropriable materials
   - Two way knowledge sharing
   - Creating demand for the service
4. Build strong partnerships
   - Incentivizing partners with complementary strengths
   - Working with infomediaries
   - Building capacity of infomediaries
5. Plan for a sustainable project
   - Planning for financial sustainability
   - Incorporating social and institutional sustainability
6. Share lessons from the project
   - Monitoring and evaluating
   - Sharing lessons with practitioners and donors

The case of the Rural-Urban Information System (SIRU) project in Cajamarca, Peru, is used to illustrate how these elements apply in practice. In rural Cajamarca, the local economy depends primarily on agricultural and dairy production, extension services have been disbanded and smallholder farmers and local producers' information needs are no longer met by the state. The SIRU project was set up to meet those information needs, by using ICTs to link local information centres (infocentres) in the region to information providers such as government bodies and NGOs working in the region. The project's successes and the challenges it faces are described in Section 6.

From this case study, we argue that project success depends on attention to process and recommend two key principles that projects should adhere to. The first is that communities need to specify the development outcomes they want to achieve, before any other choices are made e.g. about technologies or information. The second is that projects need to adopt an iterative project cycle of researching and planning, implementation, evaluation, and learning and sharing, to ensure that practitioners keep revisiting their assumptions, learning from experiences and involving the community at each stage.

Therefore we group the recommendations for best practice into a checklist, which practitioners can consult as they plan and research, implement, evaluate, and share lessons from a project. It aims to support practitioners in deciding who needs to be involved in which activities at which stages of a project. The paper concludes that best practice in connecting the first mile involves putting the last first throughout the project process. Only then can projects truly claim to have had a development impact and ICTs can be proven to support information sharing at the first mile.

1
Connecting the first mile – the challenge of putting the last first

The world that we live in is being transformed by the emergence and convergence of information communication technologies (ICTs) which make it possible for information to be processed and shared faster, cheaper, and more widely than ever before. There is a consensus among development agencies that ICTs can play an important role in development, for example by connecting people to more accurate and up to date information, equipping them with new skills, or connecting them to an international market. Trends in the ICTs industry such as cheaper hardware production or media convergence create an expectation that technology leapfrogging could level the playing field between developed and less developed countries.

Many international initiatives have been established to bring together civil society, the public and private sectors to harness ICTs for development. These include the DOT-Force (created in 2000) and the UN ICT Task Force (created in 2001). Since the publication of the World Development Report on Knowledge for Development (World Bank 1998b), development agencies have recognized the 'knowledge gaps' and 'information problems' faced in developing countries. Initiatives set up to address information and knowledge for development include the Global Knowledge Partnership (founded in 1997) and the international summit on the Information Society, hosted by the International Telecommunication Union (ITU).

Across the world, agencies are piloting projects using ICTs to improve access to information in developing countries. These projects are based on the assumption that better access to ICTs leads to better information, which leads to better knowledge and decision making, which leads to development outcomes such as improved livelihoods. They face the challenge of sharing information with people who have little experience of ICTs, low levels of literacy, little time or money, and highly contextualized knowledge and language requirements. For many years this was characterized as the 'last mile problem', a term borrowed from the telecommunications field, where it referred to difficulties of improving connectivity in remote rural areas. In recent years, this 'problem' has been re-conceptualized in terms of 'connecting the first mile' which privileges the needs of people living in those remote areas (Paisley & Richardson 1998).

The evidence from a first generation of projects suggests that many are not succeeding in connecting the first mile because they are top-down, rather than participatory (Gumicio Dagron 2001; Stoll et al. 2001b); they stress the role of technology rather than social development outcomes (Heeks 2002) and they privilege global, scientific information over local knowledge (Ballantyne 2002; Chapman, Slaymaker & Young 2003). Connecting the first mile goes beyond 'bridging the digital divide' or providing 'universal access' to the internet in less developed countries – it requires a detailed understanding of how information can contribute to development outcomes in the local context.

This paper aims to answer the question 'What is best practice in connecting the first mile?' through an analysis of the literature and a case study based on practical experience of an ICT for development project in Cajamarca, Peru and to offer recommendations for practitioners and suggestions for further research.

2

Information and knowledge for development

Communication has always been a valuable tool for development. Communications models have evolved from a simplistic stimulus-response or sender-message-receiver model to an understanding of the complexities of social networks and the interactivity of communication (Hovland 2003a). Techniques such as social marketing, advocacy, education-entertainment, participatory strategies and social mobilization have been in use for many years (Waisbord 2003). Whereas social marketing and education-entertainment place expectation on individual behaviour change, methods such as advocacy and social mobilization aim to change the structures that underpin society.

Recently, however, information and knowledge themselves have become central in the discourse on development. The World Bank report of 1998–9 makes an explicit link between knowledge and development outcomes:

> *Knowledge is like light. Weightless and intangible, it can easily travel the world, enlightening the lives of people everywhere. Yet billions of people still live in the darkness of poverty – unnecessarily. Knowledge about how to treat such a simple ailment as diarrhoea has existed for centuries – but millions of children continue to die from it because their parents do not know how to save them. Poor countries – and poor people – differ from rich ones not only because they have less capital, but because they have less knowledge. Knowledge is often costly to create, and that is why much of it is created in industrial countries. [. . .] Knowledge also illuminates every economic transaction, revealing preferences, giving clarity to exchanges, informing markets. And it is lack of knowledge that causes markets to collapse, or never come into being.*
>
> *(World Bank 1998b:1)*

The report highlights two types of problems faced by developing countries: *'knowledge gaps'* which describes the unequal access to knowledge about technology or know-how, and *'information problems'* which describes the incomplete knowledge about attributes which can lead to market failures. To narrow knowledge gaps, the report recommends that developing countries acquire knowledge by tapping and adapting knowledge available elsewhere and building on local knowledge; absorb knowledge by supporting education and lifelong learning; and communicate knowledge by taking advantage of new ICTs. To address information problems, the report suggests solutions such as processing financial information, increasing knowledge of the environment, and addressing information problems that hurt the poor.

'Knowledge-based development' (Ludin 2003) has come under fire in a context of:

> *[. . .] globalisation, the internationalisation of the trade in educational services, the aggressive marketing of Northern higher education, the continuing challenges to research in the South, and the continuing reduction in overall aid to the South.*
>
> *(King & McGrath 2003:17)*

The re-emergence of information and knowledge as central to development is also due to the influence of a powerful ICT industry shaping our concepts of development, as well as to the discourse of knowledge management and knowledge-based innovation currently influencing thought in the corporate sector. This paper argues that information can only play a role in development if people can appropriate, internalize, and use it, making it part of their knowledge.

Theories of knowledge suggest that sharing information is different to sharing knowledge. For Davenport and Prusak (1998), knowledge exists in people:

> *Knowledge is a fluid mix of framed experience, values, contextual information, and expert insight that provides a framework for evaluating and incorporating new experiences and information. It originates and is applied in the minds of knowers.*
>
> *(Davenport & Prusak 1998:5)*

For Devlin (1999), information becomes knowledge when a person internalizes it to a degree that it is available for immediate use. Knowledge exists in an individual person's mind, whereas information exists in the collective mind of a society. As Nonaka and Takeuchi (1995) point out, both information and knowledge are context-specific and relational in that they depend on the situation and are created dynamically in social interactions between people.

Knowledge management literature highlights the importance of context and of the individual in internalizing information. Critiques of knowledge sharing initiatives in development suggest that they confuse access to information with access to knowledge. For example, van der Velden's (2002) analysis of the Development Gateway suggests that the project was designed without a concept of knowledge as contextually-defined and therefore does not address the needs of key audiences:

> *The critique of the Bank's approach in this case indicates that knowledge needs to be presented in the appropriate context and be meaningful in the local situation in order to be useful and effective.*
>
> *(van der Velden 2002: 31)*

ITDG has conducted practical research into knowledge sharing for development and has found that in many cases, access to information is not access to knowledge. For example, research into the Knowledge and Information Systems of the Urban Poor concluded that, although information can be available from local government offices, the staff can act as 'gatekeepers' and refuse to share the information (Schilderman 2002). Context plays a key role in how useful information is. For instance, in an assessment of WorldSpace radio, Aley et al. (2003a) found that:

> *A common request is for more information that is appropriate to their specific context, meaning it must be locally relevant and applicable. Many people prefer information to be exchanged orally in their own mother tongue, and appreciate practical face to face demonstrations and follow-up*
>
> *(Aley 2003a:5)*

This echoes the findings of another ITDG investigation into Making Knowledge Networks Work for the Poor (Lloyd Laney 2003c), which concluded that face to face communication is the most effective mode of transferring information and that the role of the information intermediary was key to information systems facing the challenge of sharing information in a local context.

Information sharing in itself is therefore not necessarily going to lead to development outcomes, without the processes in place to support the transformation of information into knowledge. This particularly affects ICT-based information systems, which are the focus of this paper, because the characteristics of ICTs that make them useful for information dissemination can also depersonalize the information (for example covering large geographical distances or many-many communication). Section 3 examines the literature on ICT projects and specifically on projects connecting the first mile to isolate best practice in using ICTs for information sharing in development.

3
The role of ICTs in information sharing for development

Developments in the ICT industry, such as the convergence of media, faster processing, higher bandwidth, and lowering costs, have led to high expectations about the potential of ICTs to support development.

In 2001, the OECD published a study of donor ICT initiatives (Hammond 2001), which stated that the main ICT initiatives of multilateral and bilateral donor agencies amounted to at most $500 million annually, much on ICT strategies, capacity building and infrastructure. Harris (2004) cites several uses of ICTs in for poverty alleviation:

- Distributing locally relevant information
- Targeting disadvantaged and marginalized groups
- Promoting local entrepreneurship
- Improving poor people's health
- Strengthening education
- Promoting trade and e-commerce
- Supporting good governance
- Building capacity and capability
- Enriching culture
- Supporting agriculture
- Reinforcing social mobilization

In recent years, many studies have been published, particularly by practitioners in the development field, on the use of ICTs for development. A review of the literature highlights key concepts around which practitioners are polarized. An understanding of the different perspectives is essential to illustrate the different meanings of 'best practice' in discourse on ICTs and to situate this study against an ideological background. Commentators have attempted to divide the literature on development into different factions: sceptics and pragmatists, optimists and pessimists, globalists and localists. In this paper, it is suggested that authors diverge along four dimensions:

- Top down vs. participatory solutions to development problems
- Global vs. local solutions to development problems
- Technological vs. social solutions to development problems
- Optimism vs. pessimism about the role of ICTs in development

These different perspectives shape authors' discourse and priorities, their concept of 'impact', and their understanding of 'best practice'. The clash of perspectives is illustrated for example in the terminological debate over the *last mile problem*. For some authors the term typified a top-down approach to development, viewing the delivery of technologies to people living in remote areas as a solution to development problems (Paisley & Richardson 1998). In contrast, the concept of *connecting the first mile* starts from the needs of marginal communities and marks the connection between the local context and global information systems.

3.1 Top-down or participatory solutions?

In many cases ICT for development projects are driven by the donor agenda, which has a short term horizon. Projects may not recognize or be able to address the multi-dimensional causes of poverty, due to a narrow focus on donor objectives (Stoll et al. 2001a). Gumicio Dagron (2001) is especially critical of the role of donors and their focus on large scale projects:

> *The international donor community is still reluctant to acknowledge 30 or 40 years of failures and millions down the drain because of ill-planned macro programmes. The eagerness to go fast, to show short-term results, and to extend coverage to large numbers of people has actually backfired.*
>
> *(Gumicio Dagron 2001:11)*

For him, donors' concern with scale serves to multiply models that clash with culture and tradition and paralyse communication (being imposed from the outside), instead of linking communities and facilitating exchanges.

ICT for development projects are frequently criticized for failing to build on existing systems or work in a participatory way. Critics argue that 'top down' projects driven by the donor agenda, fail to achieve local ownership (Aley 2003; Gumicio Dagron 2001; Lloyd Laney 2003c) and are therefore socially unsustainable. Authors concerned with participation tend to identify it as the locus of social sustainability, arguing that the active involvement of users minimizes social exclusion and perpetuates support for the project:

> *The concept of establishing a dialogue with beneficiaries all along the process of conceiving, planning, implementing and evaluating a project has been gradually consolidating. At first, implementers understood that beneficiaries should be involved in the activities leading to social and economic development of a community, for the purpose of building up a sense of ownership within the community. This was at last perceived as important especially in terms of the sustainability of the project once the external inputs ended.*
>
> *(Gumicio Dagron 2001: 10)*

There is evidence from projects that local participation is necessary for the success of a project. There are reports of telecentres in South Africa where donated equipment is lying unused (Nicol 2003) because there is no local ownership, and also of projects which have been very successful, such as the Deccan Development Society, which adopted a bottom-up approach in its work with Dalit women on water harvesting.

3.2 Global or local solutions?

Many projects view information as a global public good, which is undersupplied by the private sector (Barnard 2003; Meyer 1997) as Stiglitz (1999) describes:

> *Disembodied knowledge has the characteristics of a public good (non-rivalrous and, once public, non-excludable) while money is the quintessential private good. [. . .] disembodied knowledge for development is indeed a global public good and, like other public goods, it would be undersupplied if left entirely to private initiative. The internet has in practice brought knowledge access closer to the ideal of a global public good. The communication revolution has made great strides in facilitating communication within*

countries and has also enhanced the ability of developing and transitional countries to tap into the global pool of (codified) knowledge.

(Stiglitz 1999:6)

The concept of a 'global pool' of information (since knowledge only resides in people) does not take account of the need to contextualize information to transform it into knowledge. Nor does it acknowledge the importance of information local to developing countries. It creates a knowledge economy where global (or Western) knowledge is perceived as more valuable. Schech describes the World Bank's 'self appointment as the manager in the creation, transfer, and management of knowledge' as another step in establishing the hegemony of Western knowledge:

It continues the dominant trend in development studies to construct the West as holding the key to the development of the South – first capital and technology, now ideas and knowledge.

(Schech 2002:19)

Ballantyne (2002) highlights the emphasis on external content 'pushed' at people living in poverty and the struggle faced by efforts to push local content (e.g. research conducted in the South or Southern arts) onto a global stage. Many initiatives offer one-way transfer of information (usually from the global to the local level) but failing to promote genuine two-way knowledge sharing:

Community knowledge partnerships that can develop mechanisms to deal with the problems of connectivity and information literacy, and incorporate local and external knowledge, can directly benefit poor people. This approach could replace the traditional process of a 'one-way' flow of information from a scientific, information rich core to a remote information poor community, with dynamic information sharing partnerships with a two-way flow of information at every level.

(Chapman et al. 2003: viii)

In practice, many ICT projects have been criticized for failing to acknowledge the local context and disseminating supply-driven information that is not relevant to that context or in the wrong language, which does not meet the local demand for information (Gomez & Casadiego 2002). This is particularly the case when ICTs are used for one-way broadcasting of information, instead of developing two-way flows.

3.3   Technological or social solutions?

Some authors raise concerns that technology is increasingly determining the solutions sought to development problems (Boyle 2002b):

What I am concerned with is the degree to which complex social development goals become seen as functional or technical problems when ICTs are introduced and how technology becomes particularly determinant in how larger goals are understood and acted upon.

(Boyle 2002: 102)

This 'technological determinism' can be ascribed to pressure from the ICT industry, to the pressure from politicians for quick solutions and to the lack of alternatives sought to the Western development paradigm (Heeks 2002). Examples in the literature cite particular types of technologies and applications, such as Wireless Fidelity (WiFi), open source software, low-cost devices, and translation engines, as 'stepping stones' towards 'digital inclusion' (Primo Braga, Daly & Sareen 2003). They recognize the difficulties of communicating to the local context (such as language or access barriers), but locate the potential to overcome them in new technologies, rather than in social factors. This can be a dangerous place to begin, since ICT tools themselves are cultural artefacts in which dominant interests are reflected (Gigler 2004), so a *'design-reality gap'* exists between the context in which a technology was created and the context of its use (Heeks 2002). For example, Batchelor et al. (2003) highlight the use of English language technology as a hindrance to the success of the Uganda Development Services.

Practitioners would benefit from adopting a structurational approach to ICTs and society (Orlikowski 2000), recognizing that each influences the other as technology is enacted in a particular context and therefore paying close attention to the cultural assumptions implicit in the technology and also in the local context.

3.4   Optimism or pessimism?

Some authors are extremely optimistic about the contribution that ICTs can make to development goals. Heeks (2002) comments on the 'current prevalence of positive and technologically deterministic viewpoints' which he sees as influenced by the climate among international development agencies:

> *A number of factors among agency staff may explain the emergence of this viewpoint. They include naivety about ICTs, desire for career advancement, pressure from ICT vendors, a lack of alternatives to the trends/fads of the Northern private sector, and pressure from political masters for quick solutions to development problems. The viewpoint also emanates from those seeking funds or guidance from the development agencies. They tend to mimic the views and messages of those agencies.*
>
> *(Heeks 2002:4)*

In this paper, we adopt the perspective that the contribution of ICTs to development goals depends on how they are used. For projects connecting the first mile, ICTs will need to support knowledge sharing at grassroots and lead to development outcomes such as improved livelihoods in order to be successful.

4

ICTs and the first mile

The field of ICTs for development covers much more than simply connecting the first mile. For example, studies are published on business applications of ICTs, software development and the software industry, outsourcing, ICT policy and strategy, ICT diffusion, e-finance, and teledensity. There are also articles on the use of ICTs for communication between development professionals, for practitioner networks, health professionals, for online activism, or for humanitarian assistance.

The literature about connecting the first mile focuses on the use of ICTs to enhance information sharing with communities at grassroots. The issues practitioners face repeatedly is how the project has contributed to development goals, how to monitor and evaluate that contribution, how to ensure the project is sustainable, and how to 'scale up' or replicate a successful project in a different context.

4.1   Impact at the first mile

Practitioners and academics struggle to define a causal link between development outcomes at grassroots and ICT projects. Initially development agencies aimed for 'universal access', seeing an 'indisputable link' between ICTs and poverty reduction (Flor 2001), so impact could be measured in terms of the number of people accessing ICTs. But practitioners now recognize that access to ICTs does not necessarily lead to development outcomes and are developing different ways of conceptualizing how ICTs contribute to development goals at the first mile, as outlined below.

Improving livelihoods and promoting opportunity

The Sustainable Livelihoods approach to development focuses on people and their capabilities. The approach seeks to understand people's strengths, including their skills and possessions, and how they use these assets to improve the quality of their lives. It considers the context in which people live and the risks they face, their livelihood assets, and the structures and processes that have an impact on their livelihoods. The aim is to develop livelihood strategies to achieve better outcomes for people.

Projects connecting the first mile aim to improve livelihoods at the local level in a variety of ways. These can include: offering access to market information; cutting out the middleman in market transactions ('disintermediation'); and offering information about grants available to producers or techniques to improve production. Op de Coul (2003) cites an example of disintermediation from Central America:

Agronegocios in El Salvador helps farmers to become traders as well and to establish direct contacts with buyers, instead of selling to middlemen (called "coyotes"). This is done through bi-weekly markets in the capital but also through a virtual market on the website where offers and demands are published. In Agronegocios centres spread around the country the farmers and their children are taught how to enter their offers and how to find possible buyers. Though the farmers in general prefer personal contacts with their customers, the virtual market has the advantage of offering "business to business" opportunities and bigger quantities can be sold. Furthermore, trade is not restricted to the province or country the farmers live in; deals with foreign traders are an option as well."

(Op de Coul 2003: 7)

In the literature on connecting the last mile, some authors review case studies according to their contribution to improved livelihoods, such as Chapman et al. (2003) and Batchelor et al. (2003), who attempt to map sustainability outcomes using the Sustainable Livelihoods Framework.

Facilitating empowerment

The World Development Report of 2000/1 (2001) identified three priority areas for reducing poverty: promoting opportunity, facilitating empowerment, and enhancing security. Enhancing empowerment in this report refers to state and social institutions being responsive and accountable to poor people, strengthening the participation of poor people in political processes and decision making, and removing the social and institutional barriers that result from distinctions of gender, ethnicity, and social status.

In projects connecting the first mile, ICTs have had a significant impact on amplifying the voices of people living in poverty and increasing participation in political processes. Many of the most inspiring ICTs projects have involved local appropriation of ICTs such as community radio or video, which have empowered communities to make a political impact. For example, ITDG's Women's Voices project (ITDG 2003) trained women's groups in the slums of Nairobi in using video so they could communicate directly to policy-makers about their situation and development priorities. The videos were shown to an audience of government ministers, housing directors, donors, and NGOs. Later the videos were shown on national television and won an international award, the Betinho Award for Technology and Social Justice. The women gained confidence and made contacts regionally and now have plans for setting up a local resource centre with access to information on tenure, health, training, and job opportunities.

In the literature on connecting the first mile, authors review case studies according to the priority areas listed in the World Development Report 2000/1 (Cecchini & Scott 2003; Op de Coul 2003). These studies group examples of projects according to the three priority areas, but in some instances little evidence is presented that, for example, increased empowerment has led to poverty reduction.

Enhancing security

The World Bank report 2000/1 defines enhancing security through an understanding of the context in which people live and the risks they may be vulnerable to, such as economic shocks, natural disasters, conflicts, etc. The report recommends risk mitigation strategies such as insurance and tackling epidemics such as AIDS.

The role of ICTs in early warning systems and the use of Geographical Information Systems (GIS) are two applications of ICTs at the first mile. For example, Batchelor et al. (2003) describe the MIGIS project, which uses GIS software in participatory processes to help people model the geographical risks in a region of China.

Preserving indigenous knowledge

The World Bank published a report in 1998 (World Bank 1998a) on the role of indigenous knowledge in development, which argued that indigenous knowledge (IK) needed to play a greater role in development activities. It suggested that sharing IK could help reduce poverty by enhancing cross-cultural understanding and increasing responsiveness of development initiatives. It argues that development agencies should disseminate information, facilitate the exchange of IK among developing country communities, apply IK in the development process, and build partnerships. The report recognizes the importance of addressing key issues such as intellectual property rights (IPR), national policies on knowledge for development, the role of ICTs in enabling the exchange of IK, and the controversial aspects of transfer relating to knowledge, information, context, and culture.

Harris (2004) offers the example of the HoneyBee Network in India:

> *The Honey Bee network in India collects local innovations, inventions and remedies, stores them on-line, and helps owners obtain incomes from local patents and commercialization of inventions. The database contains more than 1,300 innovations.*
>
> *(Harris 2004:18)*

Other networks such as the Open Knowledge Network are collecting, sharing, and disseminating local knowledge, supported by ICT-based solutions.

4.2 Monitoring and evaluation

The monitoring and evaluation of communication for development has generated a great deal of debate. In January 2005, the World Bank held an e-conference on the topic and practitioners discussed monitoring and evaluation at length, focusing in particular on the importance of monitoring of communication for development, what to evaluate, proxy indicators, quantitative versus qualitative methods, and short and long term approaches.

Attempts to develop evaluation frameworks for ICT projects in development have highlighted the current shortcomings of project evaluation. For Stoll et al. (2001) the project approach is characterized by specificity of objectives, predetermination of outputs, short term horizons, the pre-eminence of the sponsors, and intermediation of the end users' priorities by the development agencies.

It is clear from the way projects are implemented that power relations inherent in development projects affect how impact is reported (Gumicio Dagron 2001). Practitioners need to demonstrate success to receive further funding, donor organizations are unwilling to invest in monitoring activities by donor organizations and 'beneficiaries' have relatively little input into the monitoring and evaluation process. Factors such as these suggest that demonstrating results can be more important than demonstrating a development impact for beneficiaries.

4.3 Sustainability

The emphasis on sustainability in projects connecting the first mile partly reflects donor concern with short-term investments, but also the belief that sustainability represents demand for a service. Batchelor et al. (2003) evaluated projects according to their economic sustainability, social sustainability, and institutional sustainability. In their definitions, economic sustainability is achieved when a given level of expenditure can be maintained over time, social sustainability is achieved when social exclusion is minimized and social equity maximized, and institutional sustainability is achieved when prevailing structures and processes have the capacity to perform their functions over the long term.

Economic sustainability is seen by some as a key indicator of the success of a project because it is seen to reflect a genuine demand for that service. At the same time, in many development projects, donors are funding information dissemination as a public good, as Tschang et al. (2002) comment:

> *The nature of telecentre sustainability is complicated by the point that it may initially be a public good, especially in disadvantaged areas, yet must be ultimately self-supporting.*
> *(Tschang et al. 2002: 130)*

A great deal of research has been published on economic sustainability, in particular with regard to access initiatives such as telecentres or information kiosks, which have high set-up and maintenance costs and customers with little spare cash. The complicated objectives of providing information services as a public good and making them self-supporting have proved extremely difficult to reconcile and few initiatives have succeeded in covering their costs, even if they have developed viable charging mechanisms (Batchelor et al. 2003).

Social sustainability, as outlined in section 3.1 above, depends on participation and a project being embedded in local social structures.

Institutional sustainability is primarily a question of resources and capacity building, amongst project staff and partners, empowering those institutions to take control in local development issues.

4.4 Scaling up

A recognized problem with ICT for development is that most initiatives are pilot projects, which are then not scaled up into programmes. Weigel et al. (2004) comment that the changing role of government, and the capacity of the private sector and civil society to finance and implement projects provide fertile ground for beginning large scale implementation of multi sector partnerships in ICT for development and moving away from pilot projects. At the same time, the warnings from Gumicio Dagron (2001) suggest that concern with scale can lead to more failures because contexts are so different.

Increasingly, practitioners argue that ICTs should be mainstreamed (Beardon 2004) or embedded into development projects (Hagen 2005), rather than being an aim in themselves. The goals and purpose of the project would be development goals and not viewed through the lens of technological solutions.

5
Best practice in connecting the first mile

This paper focuses on the lessons learned by existing projects about best practice in connecting the first mile. Since there is not a coherent body of literature about connecting the first mile, selection criteria were identified to filter the available case studies:

- Is there an information sharing and development focus to the ICT project?
  - What are the objectives of the project? Does the project focus on information sharing, not just access to ICTs?
- Are the end users of the information people at grassroots?
- Is the article a good source of information?
  - Is the author independent from the project? Is there a good level of detail?

Using these criteria, a search was conducted on the Eldis portal to identify valuable case studies. Using the search criteria, over thirty case studies were identified for analysis and key documents were identified which offered lessons from less detailed cases, as listed in Appendix 2.

Previous literature reviews in this field have commented on the promotional nature of literature, the paucity of baseline and evaluation studies to date, the relatively recent emergence of frameworks for evaluation and the emphasis on telecentre projects and literature about Africa (adeya 2002; O'Farrell, Norrish & Scott 1999). It is also clear that ICT for development projects are relatively recent and in many cases will not yet be able to demonstrate their impact (Meera, Khamtani & Rao 2004).

In the literature on connecting the first mile, best practice is described through the use of examples of successful projects, examples of unsuccessful projects, and recommendations based on project experience. Several studies have collected a group of cases and compiled guidelines for practitioners or highlighted the lessons learned from existing projects. The lessons from case studies suggest that the success of many projects is 'situated success' in the sense that the project has worked due to a particular combination of local factors such as a strong champion or good timing. But there are also processes that most authors highlight as contributing to the success of projects in all contexts. So what are the key factors that have contributed to the success of projects and which activities constitute best practice?

5.1   Start from the people, not the technology

A lesson emerging from the literature on connecting the first mile is that the projects that succeeded have worked with the community from the beginning to ensure that the information systems respond to their needs (Carman 2005; Lloyd Laney 2003c). Several authors argue that participation is key to the sustainability of an ICT project and highlight the lessons from participatory technology development about ownership and appropriation (Aley 2003b; Beardon 2004). The success factors emerging from the case studies, discussed in more detail below are:

- Starting from community priorities
- Understanding local power dynamics
- Minimizing social exclusion
- Strengthening social capital

Starting from community priorities

Projects need to start from communities' development priorities (Stoll et al. 2001) and begin with a needs assessment which could draw on PRA or RRA methodologies (Bridges.org 2004; Cecchini & Scott 2003; Lloyd Laney 2003b).

Beardon (2004) gives the example of ActionAid's Reflect ICTs project, which brings groups of people together to discuss and analyse local issues and devise action plans, using participatory techniques. The different contexts have led to very different information strategies and systems being adopted to support communities in Uganda, India, and Burundi.

Understanding local power dynamics

Practitioners need an understanding of the power dynamics at the local level (Michiels & Van Crowder 2001). An analysis of ICT stories collected by Infodev suggests, for example, that dealing with local authorities can be an issue if the project challenges their power:

> Co-operation from local government is to be taken into account in a lot of projects. Either because project initiators need authorisation from the local government to start their project, or because the local government may even be a partner [. . .] they will not co-operate to the fullest if they feel that the empowerment coming from the project will challenge their positions of power.
>
> (Infodev 2004:1)

For example, Batchelor et al. (2003) discuss the role of the city council in the Global Voices project in Kenya, the failure of which has caused problems in the community. Projects need to research and recognize the power relationships in the local context from the beginning of the project.

Minimizing social exclusion

Projects run the risk of only meeting the needs of self-selecting groups (Beardon 2004). It is important to focus on the needs of marginalized groups such as women or the disabled:

> Too often, agencies solely communicate with the more active members in a community, leaving others behind who may remain poorly informed, thus perhaps increasing their exclusion. Agencies should avoid that and may also have to specifically target groups that have difficulties accessing information or have particular information needs, such as female heads of households, the young, the disabled or ill, or the homeless.
>
> (Schilderman, 2002:48)

Op de Coul (2003) lists some projects that focus solely on the needs of women: VideoSEWA in India, Women'sNet in South Africa, NGOCC in Zambia, and Feminist Radio in Costa Rica. She also comments that only a few organizations consciously target men and women equally and quotes a trainer on the Rural Litigation and Entitlement Kendra (RLEK) project in India, training tribal nomads to use a wireless communication tool:

Men folks are moving all the time and therefore there is need to train women also because they are left behind in the houses and in case of emergency they will need the wireless.

(Op de Coul 2003:15)

Strengthening social capital

Schilderman (2002) argues that people living in poverty get their information through social networks and it is therefore important to strengthen community social capital. This can include deliberately stimulating people to undertake a joint activity or action related to particular local needs and providing a space for the community to get together and meet.

For example, the FOOD project in Chennai in India trained 100 women's self-help groups in marketing and the use of social capital, to help them improve incomes from food products, soap, etc. The project provided each group with a cell phone to facilitate contact between production and marketing groups and their customers (Batchelor & Sugden 2003).

5.2   Blend ICTs with traditional information systems

Traditional information systems are based on social networks and often on face to face interaction (Beardon 2004; Lloyd Laney 2003c; Schilderman 2002). These information systems can be unreliable and people living in poverty can be disadvantaged because they are not able to access information. At the same time, ICTs are very new information systems and people may not trust them or know how to use them and may find that they supplant the existing information systems. Good practice recommended in the literature is therefore to blend ICTs with traditional information systems and the success factors are:

- Building on existing information systems
- Choosing appropriable technologies
- Adapting to local infrastructure

Building on existing information systems

Chapman et al. (2003) make a convincing case for building on existing systems:

Many donor-driven information systems are overly ambitious, overly complex, and over-designed. They tend to overlook the fundamental organisational processes and institutional incentives that drive information use and ignore potential 'losers' who may subsequently resist implementation. [. . .] There is a danger the current focus on Internet-based information systems in developing countries will undermine rich and effective existing information networks.

(Chapman et al. 2003:vii)

Lloyd Laney (2003c) recommends that projects conduct research into existing information systems and design initiatives that build on these. For example, the RITS project in Brazil partnered with local organizations to extend the reach of ICTs to low-income communities (Batchelor & Sugden 2003).

Choosing appropriable technologies

Project lessons highlight the need for technology to be chosen that is appropriate to the context of use and appropriable by the community. For example, Batchelor et al. (2003) highlight the need for local repair and operational skills to maintain the technology. Primo Braga et al. (2003) discuss the use of low cost technologies and free or open source software at the grassroots and Badshah et al. (2004) recommend the use of innovative technologies such as the wireless connectivity technology used by the n-Logue project in India. Chapman et al. (2003) stress the need for 'realistic technologies' to be used that are appropriate to the local context and affordable. They recommend blending communications approaches, citing the example of the Kothmale project in Sri Lanka where a community radio station browses the internet at the request of listeners:

> *A combination of linking old and new technologies, use of mass media and technology sharing can reach the greatest number of people, over the largest distances and with the least infrastructure investment.*
>
> *(Chapman et al. 2003: 28)*

This is supported by Schilderman's (2002) research, which shows that successful examples of strengthening the knowledge and information systems of the urban poor are rarely based on a single method of communication and that incorporating traditional media can promote two-way knowledge sharing.

Adapting to local infrastructure

In ICT projects, problems with local telecommunications infrastructure can be 'one of the biggest challenges, especially in developing countries' (Infodev 2004). To overcome this challenge, Chapman et al. (2003) recommend analysing telecommunications and IT infrastructure deficiencies to plan for realistic measures. Several projects, such as the RLEK and N-Logue projects in India have adopted innovative technologies such as the wireless to overcome local infrastructure challenges. Other authors recommend developing more accessible devices such as the Simputer (Primo Braga, Daly & Sareen 2003).

5.3   Share information that users can appropriate

Projects connecting the first mile often assume that improved access to ICTs leads to improved access to information, which leads to improved knowledge and decision making and therefore development outcomes. Evidence from projects suggests that in many cases the information is difficult to appropriate because it is exogenous, in an inaccessible format, or not from a source people trust (Lloyd Laney 2003c). The success factors in sharing information that users can appropriate and turn into knowledge they can apply are described below:

- Researching information needs
- Developing appropriable materials
- Two way knowledge sharing
- Creating demand for the service

Researching information needs

The key to developing appropriable materials is to meet local information needs. Best practice therefore requires that practitioners work with communities to research information gaps and understand information needs (Beardon 2004), then develop materials in the right format and language for use.

For example, Batchelor et al. (2003) describe Gyandoot, the Indian information system, as developing the information services in conjunction with villagers:

> *The services/facilities on Gyandoot have been chosen through a participatory process involving the community, government officials and the Gyandoot team. During the formation of the project proposal, a detailed RRA/PRA exercise was taken up involving the villagers and the community. The selection of the services was a result of this interactive exercise and was based upon the advice and the felt needs of the villagers.*
>
> *(Batchelor et al. 2003)*

Developing appropriable materials

There are lots of toolkits on offer to practitioners that recommend approaches to designing information to the local context, paying attention to the dimensions of language, cultural context, information delivery channel, and information format (Aley, Waudo & Muchiri 2004; Fisher, Odhiambo & Cotton 2003).

MANAGE, the Indian information service for farmers, highlights the problem that there is limited access to web sites in the local languages – there is a little in Hindi and very little in Telugu (Batchelor et al. 2003).

Two way knowledge sharing

Many of the projects connecting the first mile privilege exogenous information although knowledge sharing at a local level can be more relevant to local problems (Chapman, Slaymaker & Young 2003). Increasingly, development organizations are moving away from one way knowledge transfer models towards two-way knowledge sharing. Ballantyne (2002) comments on the importance of local content in this two-way flow and suggests that best practice in facilitating local content creation includes valuing and motivating local content (through rights and incentives) and building the capacity of the target group in content creation. He also recommends that projects connect to traditional knowledge and promote local participation throughout the project.

An example in practice is the ACISAM project in El Salvador, where members of a community use video and audio to document issues on mental health and feed the outputs back to the community via loudspeakers, radio, and cable television (Batchelor et al. 2003).

Amplifying local voices

Amplifying local voices, for example by incorporating local and external knowledge into information materials and connecting the target group to policy-makers, is important for community empowerment. For example, villagers using the MANAGE information system are able to demand their rights from state officers:

> *The information access at the village level is putting pressure on the middle and senior level state officers for delivering the programmes and schemes in time and to the needy. They are also under constant pressure as transparency throughout the system has improved. The villagers know their eligibility for housing loans, crop loans and other schemes and they are able to inform the concerned officers about their demands with full supporting documents, very much in time, due to information availability through the websites.*
>
> *(Batchelor et al. 2003)*

5.4 Build strong partnerships

Projects connecting the first mile bring together a wide range of actors such as community groups, information providers, media partners whose role is to reformat information, technology partners who support the ICT hardware and software, donor organizations funding the project, and information intermediaries (infomediaries). Partners have complementary strengths and coordinating multi-stakeholder projects can be hard work. In particular, the literature stresses the role of the information intermediary (infomediary), who turns information into knowledge for local users. The key success factors arising from practical experience, described in more detail below, are:

- Incentivizing partners with complementary strengths
- Working with infomediaries
- Building capacity of infomediaries

Incentivizing partners with complementary strengths

Partnerships and institutional arrangements are given central importance in the literature. In order to deliver information services to the poor, information providers need to form strong partnerships with other information providers (Batchelor et al. 2003), organizations that promote services and raise awareness amongst end users (Cecchini & Scott 2003) as well as organizations offering the technological infrastructure and finance to keep the project afloat. So, best practice includes selecting partners with complementary strengths.

Bridges.org (2003) describe the case of the Satelife PDA project which involved a range of partners with complementary strengths. The project aim was that physicians, medical officers, and medical students test PDA devices in the context of their daily work environments. Partner organizations working on that project in Uganda included the American Red Cross, Makerere University Faculty of Medicine and HealthNet Uganda, who provided technical support and project assistance. Medical texts were obtained from Skyscape, an online information provider.

Incentivizing partners to participate in the project is crucial to success. Donor organizations are looking at public-private partnerships (Carlsson 2002) and there is a case for businesses to partner with businesses, NGOs, and community groups already established in developing country markets to minimize risk and maximize infrastructure (Prahalad & Hammond 2002).

There is also a need to clarify roles and responsibilities. For example in the CARDIN network on disaster management, a clear definition of tasks, roles, and responsibilities was necessary to ensure information was shared (Batchelor et al. 2003).

Working with infomediaries

Infomediaries help communities to find the information that they seek. Different studies define infomediaries in different ways. For Cecchini et al. (2003) they are the human intermediary between poor people and ICTs. For Lloyd Laney (2003c), they represent the 'face to face' contact which is essential in turning information into knowledge for poor people. For Raab et al. (2003) they are employed to provide information, for example by government extension systems, NGOs, academia or the private sector. For Schilderman (2002) they are 'information producers and suppliers, who do so out of duty or desire', such as the public sector, NGOs, or religious organizations.

Cecchini et al. (2003) offer examples from rural India of best practice by intermediaries:

Successful examples of ICT projects for poverty reduction are conducted by intermediaries that have the appropriate incentives and proven track record working with poor people. In Andhra Pradesh, ANMs have been working with poor villagers on a daily basis for years. SKS, the microfinance institution, adheres to a philosophy of reaching out to the poorest women in rural areas. In Gujarat, dairy cooperatives have been the best agent to target small farmers. If these intermediaries are grassroots-based and understand the potential of ICT for social change, they can be tremendously effective in promoting local ownership of ICT projects. In rural India, many telekiosks operators are young, educated, computer-savvy, and very attached to their communities. They are also extremely entrepreneurial. In the case of Gyandoot, successful telekiosk operators – besides offering e-government services – often create and manage database and work on data entry for private clients, offer PC training, provide voice, fax, copy, Internet and many other services.

(Cecchini et al. 2003: 12)

Building capacity of infomediaries

Schilderman (2002) recommends best practice for projects with regard to infomediaries, which includes developing appropriate materials, sharing good communications practice, and capacity building:

Development agencies should sensitise state institutions towards more courteous and efficient information provision and, where resources are a real constraint, aim to provide additional resources and capacity building. Where this research has shown that smaller authorities are often better at communicating with their target population, this could be an argument in seeking wider decentralisation. [. . .]

There is furthermore a need to recognise, document and share good practice in communicating with the urban poor. Whereas many infomediaries are obviously not functioning optimally, some do exist that do well or have some exemplary projects or services, but often these are not widely known.

(Schilderman 2002:49)

Infomediaries need to acquire what Ballantyne (2002) terms 'adaptation skills', for example translating information materials to suit local conditions.

5.5   Plan for a sustainable project

Donor organizations are keen that projects be financially sustainable so that funding can be withdrawn after a period of time. If a project is socially and institutionally sustainable, then it is more able to respond to local demand. Projects will need to develop an exit strategy to ensure projects continue after practitioners withdraw. The success factors in planning for a sustainable project are therefore:

- Planning for financial sustainability
- Incorporating social and institutional sustainability

Planning for financial sustainability

The issue of financial sustainability is contentious in the literature because the goal of achieving development impact through making information available as a public good is in conflict with the need for services to pay for themselves. So for some authors, economic sustainability indicates a genuine demand for a service (Prahalad & Hammond 2002), whereas others see an area where public sector or non-profits need to fill the information gap (Meyer 1997).

Economic sustainability is especially key in the literature about access initiatives such as telecentres or information kiosks, which have high set-up and maintenance costs and low income clients. Badshah et al. (2004) highlight some initiatives that have developed innovative solutions to financial sustainability:

Several projects have a self-sustaining commercial focus as the driving factor – Drishtee (India), Cabinas (Peru), Warnet (Indonesia), n-Logue (India), Telecottages (Hungary), are all based on a business model. According to Amin, one way to structure a business driven kiosk model is as a franchise and many of the successful efforts analyzed have adopted this approach.

(Badshah et al. 2004: 223)

Harris (2004) cites best practice as suggested by Best and Maclay (2002), which includes keeping costs low, focusing on core communication applications, merging existing businesses and telecentres and using Voice over IP to promote competition. Tschang et al. (2002) suggest that for telecentres returns are increased through economies of scope and scale, network externalities, vertical integration, and agglomeration. They also highlight the importance of partnerships to overcome initial costs:

The high initial investment costs in equipment and infrastructure make it difficult to base expansion plans on local owner-operators' means. In-kind investment partnerships – e.g. the Indonesian government's vocational school system's partnering with local businesses to sponsor cybercafe, and private sector assistance; or the Indian Andhra Pradesh state government's scheme to involve long distance telephone companies – may be solutions to these problems.

(Badshah et al. 2002: 130)

The involvement of the private sector in sustainable ICT projects can often reduce costs and improve service quality and efficiency (Badshah et al. 2004). The private sector is waking up to 'bottom of the pyramid' as a potential market and multinational corporations are adopting new business models and partnership models to exploit these markets, such as the shared access model popularized by the Grameen Bank (Prahalad & Hammond 2002). Whereas public initiatives can be slow to recognize services that fail to meet demand, Prahalad and Hammond (2002) suggest that through competition, multinationals are likely to bring a superior level of accountability for performance, which could benefit end users.

Best practice in developing a sustainable business model will include identifying which services are being provided as a public good and where the project could adopt a commercial model and increase returns through partnerships, in particular involving the private sector. All projects need to develop an exit strategy to ensure that projects will be sustainable after funding is withdrawn (Ballantyne, Labelle & Rudgard 2000).

Incorporating social and institutional sustainability

The sustainability of a project after implementation depends not only on economic sustainability, but also on institutional and social sustainability (Batchelor et al. 2003). For example, Op de Coul (2003) raises the high turn-over of trained technical staff in ICT for development projects as an institutional sustainability issue in projects like Afronet in Zambia. Best practice suggestions are offered by Batchelor et al. (2003):

Institutional sustainability is said to be achieved when prevailing structures and processes have the capacity to continue to perform their functions over the long term. [. . .] The studies show three mechanisms for gaining capacity: it is possible to buy-in expertise when necessary, to hire specific skilled staff or to train existing staff (or volunteers).

(Batchelor et al 2003: 17)

5.6   Share lessons from the project

Sharing findings with other development practitioners is important to communicate best practice and to understand the degree to which a project could be replicated in a different context (Cecchini & Scott 2003). Key success factors include:

- Monitoring and evaluating
- Sharing lessons with practitioners and donors

Monitoring and evaluating

Like any development project, a successful ICT for development project will have clear objectives, clearly identified target groups and realistic plans for implementation (Batchelor et al. 2003; Bridges.org 2004). An example of good practice on the ICT Stories web site is the Jharkhand-Chattisgarh Tribal Development Programme (JCTDP), an eight-year livelihood improvement and empowerment programme targeted at resource poor rural households in nine largely tribal blocks in Chattisgarh state in India, which has identified its audience and developed its project plans according to their needs.

Practitioners emphasize the need for target groups to participate throughout the project cycle, from inception through to monitoring and evaluating (Lloyd Laney 2003c). Commentators are concerned that monitoring and evaluation indicators relate to use of technology more than to the impact for example on livelihoods (Stoll et al. 2001). They recommend linking the project goals, variables and indicators to community priorities (Stoll, Menou, Camacho & Khellady 2001), critically evaluating efforts, reporting back to clients and supporters and adapting as needed (Bridges.org 2001). Best practice involves monitoring throughout the life of the project, not only once a project is completed (Lloyd Laney 2003b).

Sharing lessons with donors and practitioners

Besemer et al. (2003) make a convincing case for sharing findings with donors to influence policy and recommend donor dialogue as a best practice. Studies on advocacy suggest that establishing credibility and communication, influence and legitimacy (Crewe & Young 2002), and developing strong relationships with policymakers (Lloyd Laney 2003a) contribute to helping an organization to achieve policy change. There is evidence that the role of donor organizations can also be determinant in the success of a project:

The initiatives having less financial problems are the ones implementing online activities and the ones whose hosting organisations have good relationships with donors.

(Op de Coul 2003:11)

6

Case Study: Rural-Urban Information Systems in Cajamarca, Peru

Cajamarca is a province in the northern sierra of Peru, famed for its traditions and its natural beauty. The town of Cajamarca, nestled in a green and fertile valley, still retains colonial buildings and narrow streets, which bustle with men and women in traditional dress selling their wares. It is renowned for its history – the Inca Atahualpa was captured and imprisoned here by the Spanish conquistadors, who demanded a ransom in silver and gold and then killed him. Cajamarca's hillsides are rich in gold and there is a large mine in Yanacocha, which has brought social and environmental problems in its wake. Farmers grow wheat, maize, potatoes, beans, and cereals on the steep mountain slopes and raise sheep and cattle to produce Cajamarca's famous cheeses. In rural Cajamarca, the local economy depends primarily on agricultural and dairy production. Extension services have been disbanded and smallholder farmers' and local producers' information needs are no longer met by the state.

In 2002, the Rural-Urban Information System (SIRU) project was set up to meet those information needs, by using ICTs to link local information centres (infocentres) in the region to information providers such as government bodies and NGOs working in the region. It is an innovative project because it has a unit dedicated to creating appropriable content for use at grassroots and training the infocentre managers in information dissemination – the Centre for Processing Information (CPI).

From August to November 2004, a researcher visited the project and analysed it against the elements of the best practice framework. Desk research was undertaken and thirty-nine semi-structured interviews were held with project staff and participants in the Rural-Urban Information Systems (SIRU) project aimed at connecting the first mile, as listed in Appendix 1. Quotations from documents and interviews were recorded in a database and coded and indexed according to the elements of the best practice framework. This case study presents the research findings and outlines the next steps for the project, based on the findings.

6.1   Project background

The SIRU project developed out of a workshop among local information providers who recognized that information sharing with local communities was inefficient. Many information products were inappropriate to the local context (for example, using printed materials for illiterate audiences or highly technical language) and there was duplication of effort between organizations. ITDG had implemented a previous project in the region funded by the World Bank, which aimed at increasing local access to ICTs by installing telephones and internet access in infocentres across Cajamarca. The SIRU project proposed to link these infocentres to information from providers, which had been formatted at the Centre for Processing Information (CPI) to be usable at grassroots.

The diagram below illustrates the information flows in the project in the form of a value chain:

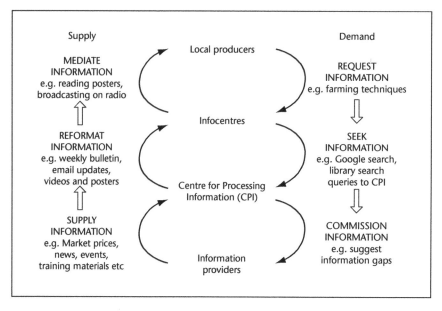

Figure 6.1 Information flows in the SIRU project.

The context of each infocentre varies dramatically – some are in highly developed urban areas and others in very rural areas, where people had a two-day walk to the nearest telephone. The infocentres in Chanta Alta, Puruay Alto, and Combayo are more rural and do not have electricity, whereas Chilete, Huanico, La Encañada, Llacanora, and San Marcos are better connected.

The map below shows the infocentres, located in rural and urban parts of Cajamarca:

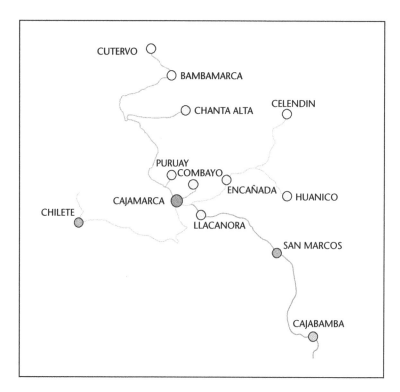

Figure 6.2 The infocentres in the SIRU project.

The other infocentres shown on the map are not yet operational. The infocentres are managed by an infocentre manager and owned by the municipality. Infocentre management is overseen by a local management committee. Most of the infocentres are in premises donated by the local government and are supported by a project partner. Project partners are from six development agencies (ITDG, SNV, CEDEPAS, CARE, PDRC-GOPA-IAK, and PRODELICA) and are responsible for providing information to the CPI and supporting infocentres through donations of equipment and staff time.

The pilot phase of the pilot began in 2002, with the selection and training of infocentre managers and production of materials by the CPI. Since then, the project has strengthened its relationships with project partners, developed many new materials and the infocentre managers have become confident and authoritative voices in their communities. The project's innovative approach has brought it recognition, in particular for the capacity building work it has undertaken with infocentre staff.

At the same time, it operates in a challenging environment: the project aims for sustainability but presently still depends on donor funds, the local contexts are varied, and socio-political factors play a key role in how much the infocentres are used. Cajamarca has high levels of illiteracy, particularly among women, who face discrimination in an extremely chauvinist society. There is little ICT penetration in the rural areas, so few among the local population have the skills to use a computer. Radio is obviously a popular medium; people can be seen in the streets carrying them around their necks.

6.2   Analysing best practice

This analysis is based on visits to five infocentres (Chilete, Puruay Alto, Llacanora, La Encañada, Chanta Alta, and Combayo). Desk research was undertaken, as well as semi-structured interviews with project staff and participants. Interviews were held with:

- the SIRU project manager
- three members of staff from the CPI
- five infocentre managers, their assistants, and two Peace Corps volunteers working at infocentres
- representatives from the six partner organizations (eight interviews in total)
- representatives from other information providers working with the project (two from the National Institute of Culture, one from the Agricultural Information Department, one from the Woman and Family Association)
- three managers from multimedia companies the CPI works with
- representatives from local organizations (three municipal authorities and the president of the Rondas Campesinas)
- three NGO field staff
- two users of the infocentre's services

Informal conversations were also held with local producers in Magdalena, Chilete.

Quotations from documents and interviews were recorded in a database and coded and indexed according to the elements of the best practice framework.

This case study structures the story of the SIRU project according to the six key aspects of the best practice framework:

- Start from the people, not the technology
- Blend ICTs with traditional information systems

- Share information that users can appropriate
- Build strong partnerships
- Plan for a sustainable project
- Share lessons from the project

6.3   Starting from the people

The purpose of the SIRU project is to offer access to relevant information for local producers, organizations, and enterprises in Cajamarca, which they could use in development processes.

Various project partners have commented that the target audience was not involved in the design of the project and participation was limited to the collection of information needs. This may explain the low explicit demand for SIRU's information services, as there is little ownership or understanding of the project among the target audience.

> [. . .] the project design has been participatory with regards to partner institutions; however the target group has not been involved; in this sense principally in the urban areas it is seen as an external project. Perhaps in rural areas appropriation is attributable to the absence of any alternative communications.
>
> (External evaluation, PRODELICA)

The power dynamics at the local level have had a critical effect on ownership of the project. In particular, the project has been affected by relationships with the municipalities, and gender and age can exclude sections of the target audience from the project.

Starting from community priorities

Participatory workshops and information mapping exercises were conducted, for example in Combayo, where the research concluded that the local information flows built on social networks, through markets and community-based organizations. External information reached people through community leaders, young people, migrants, radio and NGO fieldworkers, and was more accessible to men than women.

At the pilot stage, project staff tried to identify infomediaries who were already established at grassroots and held meetings with over fifty local organizations to build strategic partnerships. Many of these turned out to be too small to have significant impact and the project ended up working with municipalities and the existing infocentres instead. Ultimately, the design of the SIRU project was driven by partners' needs, to reduce duplication among information providers and to make best use of the existing infocentres. It has therefore been hard to integrate the information system into the social fabric of the areas. There is little local ownership of the project, particularly in urban areas, where access to information is offered by privately run cyber cafes with more, faster machines.

Power dynamics

The municipalities have contributed to the project, donating the space for the infocentre and in other cases donating machines, security, and even paying infocentre managers' salaries. However many new mayors have been very uncooperative, often trying to appropriate the SIRU resources for their own use and failing to support the infocentres and staff in their development goals. When a new mayor is elected, in many cases they try to replace existing infocentre staff (often associated with the previous mayor) with their own cronies. In extreme cases, the mayor has tried to take the machines from the infocentre, confiscated the key from the administrator or disabled the internet access so the infocentre doesn't compete with his brother's cyber café business. In these cases, infocentre managers and project partners feel helpless and the infocentre loses potential revenue because it cannot provide services. However, the municipality is the only elected representative of the local people in many of the areas where the project works, so the project has to work with them.

> *Because the building is their property, they think it's their right. They take the key to the info-centre.*
>
> *(Previous administrator)*

The more respected the infocentre managers are in their community, the more political problems they face. One commented that he had been asked to be a regidor (a local councillor), but refused because he didn't want to be accused of politicizing the infocentre. Another was asked by his community to stand for mayor, in which case he would have to give up his job in the infocentre. The SIRU project needs to decide how far the infocentres and the infocentre managers' radio programmes can be used as political platforms in these debates.

Minimizing social exclusion

The SIRU project from its beginnings aimed to incorporate a gender focus into the project. The project leader has always been a woman and both men and women were trained as infocentre managers and committee members in the project. There are only two remaining female infocentre managers, as against six men. Two female infocentre managers left their jobs after clashes with local mayors.

> *Little women [mujercitas] don't get sent to school or to the infocentre.*
> *(Fausto Villanueva, Infocentre Manager, Puruay Alto)*

The region of Cajamarca is famously chauvinist – few rural women are educated beyond primary level. It is unlikely that women under these circumstances would be comfortable asking a male administrator to explain or help her seek information. One suggestion was to work with women volunteers in the infocentre or to increase the number of women infocentre managers. The information the project provides needs to be adequate to gendered interests – on the whole women are more interested in animal rearing whereas men are more interested in market prices.

The project may also need to devise inclusion strategies to account for age. Although a wide variety of people come to use the telephone in infocentres, the internet is only really used by young people. Unless the administrator offers a special invitation to older people, they tend not to participate. One telephone user said 'It's for the school kids – teach my son.'

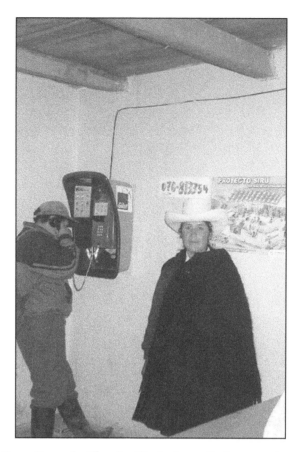

Figure 6.3 Anastasia Peres Garay in Chanta Alta believes that computers are only for schoolchildren.

The project has achieved greater inclusion of young people in decision making processes (by getting the infocentre managers into decision making fora) but it could be argued that the SIRU project has created new power relations in the area, empowering some groups and leaving others behind. Infocentre managers said that their social position had changed considerably through participating in the project – people are envious or shy with them, or sometimes come and congratulate them and offer them drinks.

Strengthening social capital

In Cajamarca there is a tradition of communal work – 'la minga'. The majority of the infocentres were offered by the municipality and did not call for much work from the community. However, in other projects where the community have participated in the building of the infocentre, there is a sense of ownership. Another good practice to create community links and promote innovation is to organize exchange visits, for example between the infocentres.

It was clear from the project that the location of the infocentre played a role in social exclusion. In the infocentres housed in municipal buildings, producers might be too intimidated to go in. A more strategic site for the infocentre might be nearer the market, where producers get together and share information every week.

It is notable that in rural areas, the local population are supportive of the infocentre and staff. For example in Puruay Alto, they organize barbecues to raise funds for the administrator. In another infocentre, when the municipality wanted to take away the computers, it was community members who stopped them. However, the infocentres do not yet seem to be somewhere the community gets together.

Figure 6.4 The infocentre in La Encañada is popular with children, who come to play board games in the afternoons.

It's not situated in the most strategic place. Few people tend to go there – even I get nervous walking in! It's not like a market or a common space.

(Luis Valera, CEDEPAS)

6.4 ICTs and traditional information systems

ICTs are very unfamiliar for many people in the rural Cajamarca, who have traditionally relied on word of mouth to get their information, particularly in the markets. They have grown confident in using the telephones in the infocentres, but using the internet requires much more skill. The project is broadcasting weekly on the local radio, a popular medium, but it is not a very interactive information source. The project faces challenges with the local telecommunications infrastructure, which are mainly due to the installation of the infocentres through a previous project.

Existing information systems

The SIRU project researched traditional information systems in the pilot phase, such as sending messages with the milk van. Project staff held participatory workshops with infocentre staff and selected participants. At one infocentre, a visitor from the University of Reading came to research information flows and needs using participatory methodologies.

Traditional information systems have been strengthened through the use of rural telephony and community radio, but information intermediaries working with the target audiences have tended not to use the infocentres, preferring their own information sources (books, colleagues, and accessing the internet from their offices in Cajamarca). The project's technical infrastructure needs attention and in particular, the project needs to acquire local technical expertise to repair equipment without delay.

Appropriable technologies

The service in greatest demand in the rural infocentres is the telephone. Using a phone locally represents a saving for the user, who previously would have to travel to a town like Cajamarca to use the phone. In Combayo, the demand is so high that the administrator regularly sells out of phone cards before the week finishes.

> *Before, people would go to Cajamarca – 3 soles each way – people who live on 10 soles a week. They go to the infocentre and call – they spend one sol and someone can call them back. So it helps them.*
>
> *(Segundo Chunque Banda, Infocentre Manager, Combayo)*

The infocentre managers have been trained in radio production and on their weekly radio programme, which can be heard across Cajamarca, they discuss and explain the contents of the SIRU manuals, interspersed with music and chat. Conversations with producers in Magdalena suggest that they listen to that radio station (Radio Layzon), but don't necessarily catch that programme.

ICTs are very unfamiliar to some of the target audiences in this region. For example, when the infocentre in Puruay was opened, the administrator had to dial the telephone number and show users which end of the telephone to talk into. Now, users have adopted the technology and can use it themselves. Many of the local people are still scared to use the computer, afraid that they might break it if they touch it. To overcome this, the infocentre administrator at Chanta Alta borrowed typewriters from all the local organizations (e.g. the municipality, the police, etc) and taught people to type before they began to learn how to use the computer. This year, the SIRU project has received some money from the FAO 1% Fund, which will be used to equip the infocentres with whiteboards and megaphones, and where there is electricity, with videos and DVD players. The infocentre managers believe these communications technologies will be more appropriate to the local context.

> *I had to teach them to type [. . .] Holding their hands – they thought they might break the machine.*
>
> *(Antenor Alva, Infocentre Manager, Chanta Alta)*

At the infocentres, if there are technical problems with the infrastructure, the infocentre managers cannot repair them themselves. They have to report them to Gilat (an local telecoms installing company) who take their time in repairing them. The infocentre in Chanta Alta waited three months for the satellite to be repaired. If equipment donated by partners or the municipality, such as monitors or printers, needs repairing, that can also take a long time. The CPI does not really have technically expert staff – when the server went down, they had to send it to another NGO in Lima to repair it. This lack of technical skills could be considered a risk for the sustainability of the project.

Local infrastructure

The SIRU project has had various problems with the infrastructure design, arising from the design of the infocentres in the initial World Bank project and the agreements with the telecom provider. In the infocentres, which don't have electricity, they depend on solar panels. In those infocentres, there is only access to the internet for two hours a day, which makes it impossible to offer a consistent service. For example, in the Puruay infocentre, the users are schoolchildren, but the internet is only available when they have left for the day. Even in places where there is electricity, the machines are very slow and compare unfavourably with cyber cafes. Two options for the SIRU project are to renegotiate the terms of agreement with the infrastructure providers or to partner with an energy programme, for example at ITDG, to install energy solutions.

The computers and server at the CPI crash regularly. The DGIA has offered space on their server to host the SIRU portal. This might solve the problem of the server crashing when a lot of data is entered to the content management system. The radio kit left over from the Chilala project has problems with the transmitters, which has caused a popular radio programme to stop broadcasting.

Three of the infocentres, Llacanora, Combayo, and San Marcos, were broken into and equipment was stolen. Where the equipment was insured, the machines have been replaced, but the San Marcos infocentre has been closed since February because the equipment was not replaced.

Figure 6.5 People in Chanta Alta are afraid of breaking the computer and the infocentre manager has to hold their hands.

Figure 6.6 Combayo infocentre has solar panels to generate electricity.

6.5 Appropriable information

The literature on ICTs for development stresses the importance of information that is appropriate to and appropriable by local people. The innovation within the SIRU project is that the CPI exists to 'reformat' or 'repurpose' information for these audiences. The project has demonstrated good practice in creating and validating appropriate materials, but has not connected demand from the target audience with the process of commissioning materials (which is driven by partners). The project is encouraging the creation of local content, although there is still an emphasis on exogenous information.

> *Not all the information is useful. What they see as necessary isn't always what we see as useful.*
> *(Alfredo Flores, CEDEPAS)*

Researching information needs

In the pilot phase of the SIRU project, a needs analysis was conducted at each infocentre to identify relevant topics. Every year, the project runs competitions for the infocentres to collect and document local knowledge. The materials produced are then distributed by the infocentre network.

Grape growers in Magdalena described the problems they have had in obtaining chemicals for their crops – although the shops or markets will sell a product, producers have no guarantee that it is what it says on the label. This is probably not information that the project can offer, although it highlights important information problems for the farmers. Many of these information problems would be addressed if the farmers were to organize themselves and demand their rights – perhaps the SIRU project could partner with agencies who promote local organization.

It was important to the end users that the source of the information be trustworthy. This is also detailed in the literature (Lloyd Laney 2003b). For different audiences, trustworthy will mean different things. Farmers may want to see with their own eyes the yield a new technique offers (e.g. at agricultural fairs, demonstration centres, or through associations). A known NGO field worker is a more trusted source of market prices than a web site distributing information collected by the Ministry of Agriculture. In fact, the System for Rural Information in Arequipa (SIRA) has taken over distributing local information from the Ministry because it is a source people trust, unlike the Ministry. Sources viewed with suspicion, like the Yanacocha mine, for example, are producing information about livelihoods, but it is unlikely to be used by local people because they don't trust the brand. NGO fieldworkers want to know the basis on which evidence has been collected, e.g. whether market prices are what consumers are paying or what middle-men are paying.

Developing appropriable materials

The project has validated its information materials with focus groups at infocentres. In the focus groups, members of the target audience demonstrated their preferences for particular information formats, such as short videos. They commented on the cultural legibility of the images used – for example, on one poster, the men were dressed in colours that only women wore. There is a highly local dimension to language use – for example a seed potato is called different names in different parts of Cajamarca: 'ambulco', 'cocon', 'lunta', 'bellota'.

Information is highly localized – one infocentre administrator said that the information about honey production wasn't of interest to local people because it was not from the area. To deal with this, the SIRU project now only employs editorial staff from Cajamarca, who are more used to producing materials for this audience. The project is conforming to good practice by validating its materials, but it could make better use of the enquiries service to find out more about demand from the target audience and could validate materials when they are being produced, instead of once a year.

Two way knowledge sharing

The information the SIRU project disseminates focuses narrowly on livelihoods activities and not on the lives of people in poverty. For some interviewees, this limited focus made the SIRU project an unattractive information source. Although there is an enquiries service that allows target audiences to seek information from the CPI and project partners, the underlying assumption is that the CPI and project partners will not consult audiences for information. This year the project is trying to correct this by working more closely with target audiences through workshops. The project is considering incorporating ICT training into more general livelihoods issues (e.g. guinea pig farming) – evidence from the Reflect ICTs project run by ActionAid (Beardon 2004) suggests this approach could be fruitful.

The SIRU project has motivated the creation of local content through competitions. For example, there was a competition in which the infocentre staff investigated local techniques such as basket weaving or honey production and worked jointly with the producers and the CPI to produce booklets, which are now distributed through each infocentre. This is the main way the infocentres feed local knowledge into information flows in the project. They also instil pride in local knowledge through murals featuring local traditions and showing SIRU videos about successful local producers. The administrator in Chanta Alta has shown a lot

Figure 6.7 Simple information materials at the Puruay infocentre, produced by the project and the infocentre manager.

of initiative in reviving local traditions through his band that promotes local music and dance and his paintings. He also revived the Rondas Campesinas in Chanta Alta. The Rondas Campesinas are local peasants' organizations who get together to defend their land and interests. In Chanta Alta, the tradition has died out until Antenor's intervention. Now, local people are organizing themselves socially, strengthening social networks and their ability to demand their rights.

From the experiences of extensionists working in the region, it is clear that knowledge is constructed dialogically between the extensionist and the farmer – technical information becomes valuable to farmers when they understand how they can use it in their own context (e.g. learning how they can compost successfully using the few resources they have when the booklet calls for more resources). This is the key value that infomediaries add to information.

Amplifying local voices

The SIRU project tends to share exogenous knowledge (e.g. market prices or agricultural techniques) with the infocentres more than valuing and motivating local content creation. The information shared through the SIRU project is seen by some as only taking into account the productive capacity of local people and does not reflect their cultural or social realities. A researcher examining knowledge and information networks in the region commented that techniques such as artificial insemination required an understanding drawn from other contexts.

Life is more integral than agricultural production – how can they collect the learning from the Andean culture? [. . .] There's a need for innovation – to recover traditional knowledge from the productive point of view.

(Socorro Barrantes, Woman and Child Association)

In the SIRU project, as yet only the infocentre staff have been trained to create their own content, for example murals, web pages, and radio programmes. Building the capacity of the target group in content creation is a slow process – local people are time-poor, often illiterate and are not convinced of the value of their knowledge. The cost of capacity building is also off-putting – for example users of the telephone in Chanta Alta were prepared to make time for internet training, but not to pay for it. There is a need for SIRU to devise differentiated training costs to make ICT training attractive to local people. One way to achieve this might be to include ICT training in workshops about activities that improve livelihoods, such as animal rearing, to show how information can support livelihoods activities.

The CPI also works with the National Institute of Culture (INC), which could share local content with a wider audience, through web links, articles in its magazine, and events advertised in its cultural agenda. The infocentres could gather stories to revive the oral tradition (which tends to be lost once electrification comes to the village). It is important to bear in mind, however, that capturing local stories takes a lot of time and there is a need to reciprocate. A local trainer said that in a workshop someone had said to her, 'We have always given – we have never received anything back.'

The project has been very successful in connecting the infocentre managers to decision-makers. Some have participated in World Bank policy evaluation programmes, others have worked with the local municipality on participatory budget planning, others are part of the Mesa de Concertacion (a regional inter-institutional decision making forum on development), others work with the Rondas Campesinas and another is running for mayor in his area. However, it could be argued that the target audience has still not appropriated the means of communication and their views may or may not be represented by the infocentre managers.

Figure 6.8 Fausto Villanueva Rojas shows off his campaign T-shirt. He is running for mayor at the request of people in his village.

Creating demand for the service

The project has been widely promoted to the target audience by the infocentre managers. They have put up posters in the streets, handed out fliers in the market, visited villages and schools and the municipalities. At the inauguration of the Llacanora infocentre, they used theatre to raise awareness about the project. The project is working to create demand for its services, but the target audiences, despite facing problems related to information, do not explicitly demand information services to solve them. This is attributable to several factors: the target audiences are not using the infocentres (due to a lack of skills, time, and awareness), they may not perceive the information as useful (due to the formats or content), there are few infomediaries to make the information relevant for them (due to the time demands on administrators and the lack of partnerships with other infomediaries). At present, the project recognizes that much of the information is supply-driven, responding to the needs of project partners rather than target audiences.

6.6   Partnerships

In addition to involving the eight infocentres, the municipalities that support them and the six project partners who provide information, the project also works with other partners whose role is also key, such as infrastructure providers, multimedia companies who produce information materials for the CPI, other information providers such as the INC, and local institutions that work with the infocentre or partners. The challenge for project staff is coordinating such a large network of partners and keeping them informed as to project progress. They are developing an understanding of the incentives for different partners and the importance of clarifying roles and responsibilities, particularly the role of infomediaries and the capacity building that can support their work.

Incentivizing partners with complementary strengths

For NGOs like ITDG or CEDEPAS, participating in the SIRU project contributes to their targets. But for private sector organizations, the main reason to participate is financial profit. It is possible that the project could achieve economies of scale by giving larger pieces of work to the design companies they work with. Also, the design companies produce information products for other NGOs that could be channelled through the SIRU project. Relations with the infrastructure providers might be improved if they had a better understanding of the project and local information needs. This year the SIRU project hopes to dedicate resources to managing partnerships.

Technology can also support partnerships and new ways of working. For example, the content management system that the SIRU project uses (ActionApplications) allows for information providers like the DGIA to add their information remotely to the SIRU portal. In the future, content syndication partnerships could be established with the SIA portal in Huaral, the Ministry of Agriculture's agrarian portal, and the profit-making SIMENA portal amongst others. They could provide content and the SIRU network could distribute that information to infocentre users.

The project needs to resolve conflicts of interest for the project partners. For example, ITDG is running a sustainable livelihoods project in Cajamarca, the Yachan project, which also has targets for information sharing. The manager is unwilling to perform joint activities because it creates difficulties in reporting to the donor, who also funds the SIRU project. Another conflict for project partners is in the production of materials – one partner won a competition and the SIRU project produced 1 000 booklets based on their content. The partner kept fifty copies and the rest were distributed through the SIRU network, when he was hoping for more copies to distribute through the NGO's other channels.

The project has demonstrated that successful partnerships depend on clear definition of roles and responsibilities. For example, in the infocentre management model, the municipality owns the premises, the administrator runs the infocentre, and a committee oversee the running of the infocentre. At the same time, each infocentre is supported by one of the project partners and the infocentre managers are supported by the CPI. In crisis situations, for example when the equipment was stolen or the mayor moved the infocentre into the municipal buildings and put his wife in charge of it, staff and partners were unclear about who should respond. In another case, there were two NGOs supporting the infocentre and the administrator was always 'being bounced' between the two. With clear terms of reference, the system would function more efficiently.

> It should be made clear what role each partner plays.
>
> (Carlos Ruiz, GOPA)

Many interviewees considered the SIRU project to be an ITDG project because it is housed in their offices and many of the infocentres were set up by ITDG under the Infodev project. The project staff are mostly employed to work in the CPI (and have relevant information processing skills) but are also required to support the infocentres in administrative issues.

Working with infomediaries

Gigler (2004) distinguishes between infomediaries who support the embedding of the technologies at the local level and the infomediaries who embed the project in the social fabric of the region. In the case of the SIRU project, the infocentre managers are the former, but do necessarily not have the networks to function as the latter. There are also more convenient infomediaries who are currently not working with the project. For example, grape producers in Magdalena can stop NGO field staff they know on their motorbikes on the way back from market towns to find out market prices.

Infocentre managers are local young people, who are recognized locally as leaders (through their participation in the project). Their role in the project is varied: they are responsible for running the infocentres and increasing revenues (to earn a living for themselves), promoting the infocentre services to the local population, distributing the information from the CPI (through workshops, murals, etc), collecting local information needs, and channelling them through the CPI. Given the local level of familiarity with computers, they are also responsible for training people in using computers, creating e-mail accounts, etc.

It is more difficult for infocentre managers to act as social infomediaries, in the way that field staff from NGOs or local organizations can, partly due to time and partly because they are not yet part of the networks where information is shared. Information therefore in many cases reaches the infocentre from the CPI, but does not reach the target audiences of farmers, local organizations, and enterprises.

There are other social infomediaries who could work more closely with the infocentres to create a local information network, such as fieldworkers from partner NGOs. These infomediaries could promote the infocentre services to the target audience, and explain the information through them. For example, when one of

Figure 6.9 Lucinda Chavez, a fieldworker for an INGO, regularly uses the infocentre at La Encañada.

the infocentre managers showed a video, the audience requested more explanation on the topic. He is now hoping to work with field staff from PRONOMACHS to run a workshop. In Chilete, there are links with a regional organization (the Coordinadora de la Cuenca del Jequetepeque), which works with farmers' associations, which the infocentre could make more of. This year, the infocentres are working more directly with farmers' associations by organizing workshops providing information on themes of local interest. The key is to provide appropriate incentives for partnership, for example reducing duplication of effort or reaching a broader audience. These infomediaries also need quality materials and training to support their information provision activities.

At the moment, the audience who is using the internet at the infocentres is mostly young people. In many cases, their parents are farmers or local producers and they could be a channel for relevant information. In many cultural projects, children play a vital role as local infomediaries, for example in competitions where they share traditional knowledge from their grandparents e.g. recipes or dyeing techniques. The infocentres could target these children as infomediaries in the future.

Although the literature suggests that entrepreneurial infomediaries give the project the best chance of sustainability, evidence from the project suggests that it is a very hard life for the infocentre managers. They do not receive a salary and live off the income from the infocentres. At the moment, in many cases this means that they have to find work elsewhere to support their families. Despite their obvious dedication to the development of their communities, they cannot dedicate themselves to the project and either have to leave the infocentre locked up or train other young people as replacements while they work. In one infocentre the users were upset because on many occasions they would be expecting a phone call and hear the phone ringing in the infocentre, but with no one to answer the call.

> *They must be paid. The infocentre managers are promoters and leaders in their communities. What do these people live on? A worker deserves pay.*
>
> *(Jose Luis Torres, PRODELICA)*

In a project like SIRU, where so much training has been invested in the individual infocentre managers, their financial situation might present a problem for sustaining the project, as they have little incentive to stay. It also endangers the information sharing aspect of the project as the services most in demand and most likely to generate firm revenues (e.g. telephone use, printing, or typing) are not related to sharing information with the target audience. This year the project is reviewing pay arrangements with infocentre managers and also building on the infocentre managers' initiative of training young people by inviting them to meetings and supporting them with training and materials.

Building capacity of infomediaries

The infocentre managers have received training in information dissemination, local communication techniques (e.g. mural newspapers and potato printing), radio production, computing, and running a small/medium enterprise (SME). They were pleased with the training but would like further computing training and to review their knowledge to date. They use the teaching materials they were taught with, such as a basic, illustrated guide to computing. The infocentre managers have been trained to search for information on Google, but might find a web directory useful.

In the infocentres, it is the infocentre managers' job to contextualize information for their area. They cut out relevant news stories and stick them on the walls of the infocentre. Where people can't read, they will read manuals out to them and explain the pictures so they can use the manuals at home. The information can be quite technical and the infocentre managers are not trained in agriculture or veterinary sciences. In some cases, being farmers themselves, they experiment with the techniques in the SIRU manuals to improve yields and become experts, but mostly they need to work with local experts, trained by NGOs.

> *People don't really like reading [. . .] I read the leaflet to them and give them the handout explaining it a little.*
>
> *(Antenor Alva, Infocentre Manager, Chanta Alta)*

In the region, NGOs often work through training up local people as 'promoters' of particular skills such as veterinary sciences or irrigation and these local experts can sell their services to local producers or exchange it for gifts in kind. The difficulty for the SIRU project is to persuade these experts to give away the knowledge that makes up their livelihoods. Other potential social infomediaries include NGO field staff who work with local producers, whose job relies heavily on sharing information. However they tend not to seek information from the infocentres, preferring their own sources such as colleagues, networks, or universities. In many cases they work for organizations that support the SIRU project, such as CARE or CEDEPAS, and could share training materials and run workshops.

The quality of information the promoters have is uneven – they need tools – they're not totally prepared.

(Nestor Fuentes, Fieldworker, ITDG)

The infocentre managers have created a network to learn from each others' experiences and advocate for their rights. A previous administrator said that the infocentre managers had to learn leadership skills and self-esteem to equip them for the demands of their positions. They have learned from each other – for example the administrator of La Encañada copied an idea from the administrator at Puruay Alto to raise funds for the infocentre. The SIRU project staff has also recognized a need for an informal but private channel for infocentre managers to share their concerns outside the formal project meetings.

There are other organizations which could function as social intermediaries, such as local municipalities or Agrarian Agencies. From interviews with DGIA and NGO staff, it was clear that these infomediaries can be uncooperative and forbidding when dealing with the target audience. According to the DGIA, the Agency staff could learn from the SIRU staff's motivation and training. Project partners described the coldness of municipal staff as creating a 'breach in communication' with local producers, which stops them from seeking further information.

There are certain fears associated with going into the municipal building and the staff are not accessible, they don't receive you well, which causes a breach in communication.

(Jose-Luis Arteaga, PRODELICA)

6.7 Sustainability

The SIRU project has taken an entrepreneurial approach to commercializing the CPI and infocentres' services.

The most popular service in rural infocentres is the telephone, but few of the target audience have the time or skills to use the computers. The project is currently assessing demand for services to decide on charging mechanisms for different contexts. The staff are also reviewing funding for the project itself, which depends too much on a single donor at present. Project planning will also take account of the skills needed in-house to sustain the project and an exit strategy to ensure the infocentres can continue to work when project funding ends.

Financial sustainability

In the infocentres, there is no charge for information, but connected services such as internet or phone usage, video rental, CD rental, borrowing booklets or books, making photocopies, typing, producing promotional materials, etc are all chargeable. The CPI charges partners for the production of videos, booklets, and web sites. There are other organizations like MICHACRA, which charge for information services.

In some infocentres they are also charging for training, e.g. charging one sol each and then training twenty people at a time. In most of the infocentres, there is only one computer so training twenty people can be inefficient. All the infocentre managers are hoping that the project or municipalities will be able to provide more

Figure 6.10 The most popular service in the Combayo infocentre is the phone, with queues forming on market days.

machines so that they can train people more effectively. There is also a business model for the radio programmes in Cajamarca – the airtime costs 30 soles and advertisers will pay up to 120 soles for a slot.

At present, the infocentres are not a profitable proposition. Many interviewees felt that infocentre staff time should be subsidized, but others argue that staff might show less creativity and have fewer incentives to turn the infocentres into profitable businesses. The infocentres could be combined with shops, bookshops, tourist centres, or handicraft shops to increase revenues, although this could diminish their development impact. Another option is to seek funding from the Ministry of Agriculture or the municipality.

The SIRU project has received financial contributions from all the project partners, towards equipment and the infocentres, but the major donor it depends on is PRODELICA, who represent the EU in the region. As such, if there are delays in receiving funds from PRODELICA, the project is paralysed. Another similar project had to wait for three years before receiving funds for their information system and was forced to apply to other donors to fund staff time. The project staff spend a great deal of time on completing donor requirements such as organizing external evaluations or attending meetings. The project planning cannot be very flexible because the project needs to report against its activities as outlined in the logical framework submitted to donors. The staff are aware that they need to find new sources of funding, for example to donate or buy new equipment for the infocentres.

> *The donors definitely have an impact – we're paralysed because of them.*
>
> *(Carol Aliaga, SIRU)*

Institutional and social sustainability

The project staff are highly motivated but lack experience and basic project infrastructure, e.g. cars to get to infocentres, working computers, etc. This and the complexities of relationships with partner organizations and funders means that staff time can be inefficiently used. The infocentres need a great deal of support and there is currently no permanent member of staff to support the infocentre managers, nor a member of staff with technical skills. In many cases, the CPI staff, who are trained in communication, double up as project staff to support the infocentres.

In the project's lifetime, various expert consultants were employed, especially to train the infocentre managers in local communication techniques, multimedia production, etc. But in some cases, it was hard for project staff to monitor the achievements of the consultants. So the project now uses in-house staff and volunteers instead, but does not have enough technical skills in-house.

An ideal scenario would be for the CPI to have the skills to support infomediaries, particularly the infocentres, by offering services such as maintaining equipment, producing materials, aggregating information, searching for information, providing databases, etc. As yet, only the infocentre managers have received training, not the project partners. In the future, training for other infomediaries such as promoters or NGO field staff could be considered to help the project reach a wider audience.

The SIRU project needs to design an exit strategy to ensure the institutional and social sustainability of the infocentres after the project is wound up. In ITDG's Yachan project, the promoters trained in irrigation or veterinary sciences are linked to SENASA, a government body, which recognizes their qualifications and pays them to share information. A similar scheme where the infocentres were recognized, for example by the Ministry of Agriculture, for their information sharing activities, might allow them to continue with their work after the project ends.

6.8 Sharing project lessons

Monitoring and evaluation

The SIRU project has a good planning and monitoring system in place which includes monthly meetings with infocentre managers, fortnightly meetings with the CPI team and regular meetings with project partners, reports to donors, and evaluations. However, it is not always possible to stick to the schedule because external factors, such as funding being received late or technical hitches, can prevent progress against targets. The project works well with end users to validate information materials but has not evaluated the services as a whole participatively.

At present, most interviewees agree that the project is not reaching its target audiences – the information reaches the infocentres and is distributed to infocentre users (who are not necessarily from the target audiences). The project is only recently beginning to collect and analyse user information (e.g. what organizations they come from, what information interests them, which services they use) so does not currently have evidence of the penetration among the target audience or social impact.

The impact on livelihoods has not been measured.

(Rocio Ara, SIRU)

In order to evaluate the project work critically, the project needs to document the lessons learned. For example the monthly reports from the CPI capture the number of enquiries answered, the information emailed to networks, the bulletins and information products produced, but do not capture the lessons learned by the project. Staff turnover on the project is high and a lot of learning has been lost. The SIRU portal is currently being redesigned with a section where project documents and evaluations can be shared with project partners.

The project staff learn from experience and try to change the project accordingly, within the confines of the activities agreed with donors. For example, when the infocentre managers wanted whiteboards to communicate information to infocentre users, staff found funding from the FAO.

Sharing lessons with practitioners and donors

The project has learned lessons from benchmarking with other information systems in the agricultural sector. Each is trying to demonstrate sustainability and demand for their information products, but it appears that a need for information may not translate into a demand for information systems. Recent projects such as the SIA in Huaral have achieved good results by working in partnership with irrigation associations, who can channel the demand for information and potentially fund the project in the longer term. There has been interest from another farmers' association in Jequetepeque, who have asked for a similar information system to the one in Huaral. This system uses sophisticated technology (WIFI, thin client, etc) and the portal has excellent information design, but is highly expensive and relatively unproven as it has just launched. It does not incorporate processes for information collection and processing, which presents a problem for its sustainability.

The SIRU project has received visits from ICT4Dev commentators across Latin America and has shared its best practice through videos and presentations at conferences both nationally and internationally at for a such as the WSIS. The team leader still views it as a pilot project, and is reluctant to suggest replicating the project until lessons have been documented and shared.

7

Lessons for practitioners

Many of the challenges the SIRU project faces are because the project was designed with little participation from the target audience. The project did not identify with them what the development outcomes were that they were aiming to achieve through improved information dissemination. There are various different expectations of the project, which are hard for project staff to reconcile, such as: offering internet services, functioning as a virtual library, training people in use of the internet, communicating NGO information to a wide audience, and managing inter-institutional knowledge.

In the future, the project will be working to address the challenges it faces by working with community groups and local infomediaries to share information and understanding demand better. It will be monitoring the usefulness of the system for target audiences and infomediaries such as NGO field staff or promoters and commissioning materials based on explicit demand for information. The staff are also considering aiming more online information services at infomediaries, removing the need for costly reformatting of materials.

This paper has provided supporting evidence that agrees with Harris (2004) that the biggest challenges facing projects connecting the first mile are not technological. The approach adopted by many practitioners is shaped by the discourse on ICTs, the power of donors, and the discourse on information and knowledge which separates knowledge from the knower (van der Velden 2002) and privileges exogenous information over local knowledge (Ballantyne 2002; Scech 2002).

Boyle (2002) has raised the concern that complex social development goals become seen as functional or technical problems when ICTs are introduced. Therefore, instead of starting from development outcomes, practitioners begin by making technology choices. This can be a dangerous place to begin, since ICT tools themselves are cultural artefacts in which dominant interests are reflected (Gigler 2004), so a 'design-reality gap' exists between the context in which a technology was created and the context of its use (Heeks 2002).

Projects tend to be top-down interventions driven by the donor agenda, which has a short term horizon and rigid project management structures. Lessons are not incorporated into the project design, because *'double loop learning'* is not taking place (Hovland 2003a; Hovland 2003b). Ownership of the project and resources rests with the practitioners, and too often 'participatory methodologies' can involve consulting a self-selecting group such as community leaders and neglecting the needs of the rest of the community. In the case of the SIRU project, although community members were consulted about their information needs at the beginning of the project, they were not involved in the project design and consequently through the project cycle, the project found it more difficult to meet their needs.

7.1   The importance of process

For many authors, obstacles to connecting the first mile are located primarily in the local and national context (e.g. language, literacy, technological skills, etc) and project success depends on addressing them. This paper argues, after Beardon (2004), that the major obstacles to overcome stem from the approach a project adopts to the end users and that attention to process determines the success of a project in dealing with these. In particular, projects need to adhere to two principles.

The first is that communities need to specify the development outcomes they want to achieve, before any other choices are made, e.g. about technologies or information. The community's priorities will determine the 'information intensity' of the project (i.e. the role information plays in achieving outcomes). This will then determine how complex the information will be, whether it will be indigenous or exogenous information, which partners and technologies will help communities digest and use the information, and how this can be funded.

The second is that projects need to adopt an iterative project cycle of researching and planning, implementation, evaluation and learning and sharing, to ensure that practitioners keep revisiting their assumptions, learning from experiences and involving the community at each stage.

These are illustrated in the diagram below:

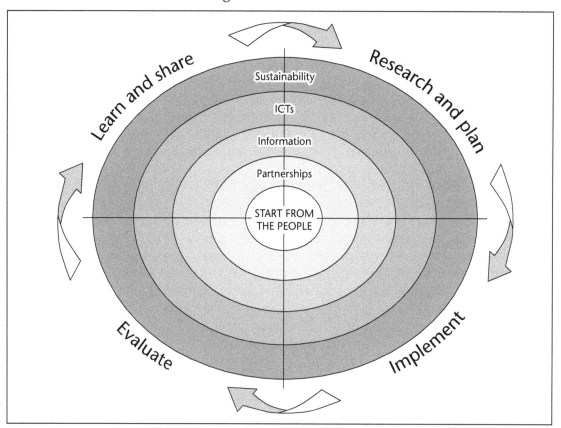

Figure 7.1 An iterative project cycle for connecting the first mile.

Based on these two principles, the next section proposes a framework to support practitioners through the project cycle, in identifying who should do what and when. The framework cannot model the richness and diversity of a project, but aims to offer a checklist of factors that have shown to be critical to success in existing projects. In particular, the framework cannot represent the iterations involved in project management. For example, within the planning stage there will be planning, implementation, evaluation, and sharing. With the framework, we do not claim to list all best practice elements but to highlight key activities and actors at each stage of the project cycle, to help practitioners pay attention to process.

7.2 Best practice framework

The elements of the best practice framework can help practitioners to deal with the challenges of connecting the first mile. The checklist below groups these elements according to the different stages of the project cycle, the choices practitioners need to make, and the actors involved:

Table 7.1

	Choices	Activities	Practitioners	Policymakers	Community groups	Technical partners	Donors	Infomediaries	Media partners
Research and plan	Start from the people, not the technology	Work with communities	✓		✓				
		Understand power dynamics at local level	✓	✓	✓			✓	
		Focus on marginalized groups	✓		✓			✓	
		Define clear objectives	✓		✓				
		Plan realistically for implementation	✓		✓				
		Link project goals to priorities	✓		✓				
	Plan for sustainability	Identify which services provided as public good and which can be commercial	✓		✓	✓		✓	
	Share information users can appropriate	Conduct a needs assessment	✓		✓				
		Identify the target group	✓		✓				
		Involve target group in project planning and design	✓		✓				
		Research information systems of target group	✓		✓			✓	
	Forge strong partnerships	Research policy environment	✓						
		Recognize power relationships	✓						
		Incentivize partners with complementary strengths	✓			✓		✓	✓
		Identify grassroots-based infomediaries with a track record of working with poor people	✓		✓			✓	
		Find entrepreneurial infomediaries	✓		✓			✓	
	Blend ICTs and existing information systems	Analyse local infrastructure	✓			✓			
		Adopt technologies that local people can repair	✓		✓	✓			
		Choose technologies that people can afford to use	✓		✓	✓			

Table 7.1 contd.

	Choices	Activities	Practitioners	Policymakers	Community groups	Technical partners	Donors	Infomediaries	Media partners
Implement	Start from the people, not the technology	Build social capital through joint activities and communal space	✓		✓			✓	
		Promote local participation in the project	✓		✓			✓	
	Plan for sustainability	Develop an exit strategy	✓	✓	✓	✓	✓	✓	✓
		Involve the private sector	✓			✓		✓	
	Share information users can appropriate	Incorporate existing systems into project	✓		✓			✓	
		Connect to traditional knowledge	✓		✓			✓	
		Address local language issues	✓		✓			✓	
		Develop materials in the right format for use	✓		✓			✓	✓
		Provide training in efficient information provision	✓		✓			✓	
		Provide useable information resources	✓		✓			✓	✓
		Value and motivate local content through rights and incentives	✓		✓			✓	✓
		Build capacity of target group in content creation	✓		✓			✓	✓
		Incorporate local and external knowledge in information materials	✓		✓			✓	✓
		Promote knowledge sharing at a local level	✓		✓			✓	
		Connect target group to policy-makers	✓	✓	✓				
	Forge strong partnerships	Provide partners with incentives	✓		✓		✓	✓	
		Negotiate conflicting interest	✓	✓	✓	✓	✓	✓	✓
		Buy in experience, hire skilled staff or train existing staff/volunteers	✓						
		Build adaptation skills, e.g. translating content to suit local conditions	✓		✓			✓	
	Blend ICTs and existing information systems	Adopt innovative technologies	✓			✓			
		Develop more accessible devices	✓			✓			
		Blend communications approaches	✓		✓			✓	✓

Table 7.1 contd.

	Choices	Activities	Practitioners	Policymakers	Community groups	Technical partners	Donors	Infomediaries	Media partners
Evaluate	Start from the people, not the technology	Evaluate efforts critically	✓		✓				
		Adapt the project in response to findings	✓		✓	✓			
		Monitor regularly	✓		✓				
Learn and share	Share information users can appropriate	Establish credibility, influence, and legitimacy with donors	✓						
		Recognize, document, and share good practice in knowledge sharing at grassroots	✓		✓			✓	
	Share project lessons	Communicate best practice to practitioners	✓						
		Share findings with donors	✓				✓		

8
Conclusions

This paper demonstrates through a review of the literature and a primary case study that the challenges facing projects connecting the first mile are not so much technological as political. These challenges are present in the local and national context and in the discourse of connecting the first mile, which shapes how projects are managed.

The paper argues that project success depends on attention to process and recommends two key principles that projects should adhere to. The first is that communities need to specify the development outcomes they want to achieve, before any other choices are made, e.g. about technologies or information. The second is that projects need to adopt an iterative project cycle of researching and planning, implementation, evaluation, and learning and sharing, to ensure that practitioners keep revisiting their assumptions, learning from experiences and involving the community at each stage.

A framework of best practice has been collated that practitioners can consult as they plan and research, implement, evaluate, and share the lessons from the project. In the future, the framework could be used to evaluate potential projects prior to funding, so projects meeting many of these criteria could be prioritized. The framework could also be used to benchmark different projects, for example by the Ministry of Agriculture in Peru to compare the various agricultural information systems in the country. The framework needs to be integrated with other tools such as an information value chain in order to understand the information flows in knowledge sharing projects at the first mile and latent and explicit demand for information.

The process of researching this working paper also offers lessons for researchers in ICTs for development. The interviews with project staff and partners and infocentre managers were conducted by a researcher who was not from Cajamarca. Due to political unrest in Cajamarca, the visits to the infocentres could only take place during one week and it was not possible to visit infocentres in Huanico or San Marcos due to time restrictions. More time was necessary to meet the target audiences, e.g. farmers, local producers, or enterprises. In future, instead of a single evaluation, a longitudinal approach would be beneficial, so the project's achievements could be chronicled over time. This would demonstrate how learning is incrementally incorporated into project planning and what the impact of the research has been on project planning.

Further research would take a set of primary case studies from various contexts and develop the framework on the basis of their experience. Adding measurable indicators would help practitioners to benchmark their projects. An analysis of the lessons of previous development initiatives such as participatory technology development or the Green Revolution could offer further suggestions for best practice.

In conclusion, the lessons from existing projects and the primary research in Peru suggest that best practice in connecting the first mile involves putting the last first throughout the project process. Only than can projects truly claim to have had a development impact and ICTs can be proven to support information sharing at the first mile. These are not new conclusions, but have yet to be enacted in the processes of development agencies. In addition to frameworks for practitioners, there is therefore a need to campaign for more inclusive ICT projects and for agencies to recognize that the challenges of connecting the first mile exist not only in the local context, but also in the way projects are managed.

APPENDIX 1

List of interviewees

- Alamiro Marcelo Gonzalez,
  Llacanora Infocentre Manager
- Alfredo Flores, CEDEPAS
- Anastasia Peres Garay,
  Telephone user, Chanta Alta
- Andres Llanos
  Rondas Campesinas, Combayo
- Anita Angulo, CEDEPAS
- Antenor Alva Ortiz
  Chanta Alta Infocentre Manager
- Carlos Ruiz, GOPA
- Carol Aliaga Rojas, SIRU
- Damian van der Heyden, SNV
- Daniel Godoy, DIA
- Eleazar Plaza Guerra
  La Encañada Infocentre Manager
- Fausto Villanueva Rojas
  Puruay Alto Infocentre Manager
- Gabriel Ynga Fernandez, SIRU
- Giovana Yactayo Ruesta
  Chilete Infocentre Manager
- Gonzalo Valdera, CEDEPAS
- Heather Van Nurden,
  Peace Corps, Llacanora
- Isabel Morales, PUBLISER
- Ismael Linares,
  Mayor of Combayo
- Jaime Soto Castaneda, SIRU

- Jorge Elliott, ITDG
- Jorge Lozano Mejia,
  Mayor of Llacanora
- José Luis Arteaga, PRODELICA
- José Luis Torres, PRODELICA
- Lucinda Chávez,
  Internet user, La Encañada
- Luis Valera, CEDEPAS
- Lorena Gonzalez, SONOVISO
- Manuel, Fernando and Guillermo
  Assistants, Puruay Alto Infocentre
- Maritza Álvarez,
  Municipal authority, Llacanora
- Marlene Serna, CARE
- Néstor Fuentes, ITDG
- Noah Ingbert,
  Peace Coros, Chanta Alta
- Norma Padilla, SONOVISO
- Ozman Altamirano, GOPA
- Rocío Ara Abanto, SIRU
- Sandra Armas,
  National Institute for Culture
- Santos Morocho
  Infocentre guard, Combayo
- Segundo Chunque Banda
  Combayo Infocentre Manager
- Socorro Barrantes, AMF
- Sol Blanco, ITDG

APPENDIX 2
Summary of cases reviewed

Case study	Summary and author
ACISAM, El Salvador	The aim of the project is to enable members of a community to acknowledge their human capacity and address common mental health problems in order to improve their social, economic, cultural, and ecological environment. The community use audio and video to capture their local problems (on mental health) and feed the outputs back to the community via loudspeakers, radio, cable television. (Batchelor et al.)
Afronet (the Inter-African Network for Human Rights and Development), Zambia	Afronet uses a web site and e-mail for the dissemination of information and communication with their stakeholders. Internally they have PC's and access to e-mail and internet to make their work more efficient. (Op de Coul)
Agronegocios, El Salvado	Agronegocios ('Agrobusiness') is an initiative directed towards training and technical assistance to small producers in El Salvador. A web site with practical information,some computer centres and technical videos are the ICT tools used to assist the farmers. (Op de Coul)
CARDIN, Caribbean	The aim of the project is to strengthen the capacity within the Caribbean community, for the collection, indexing, dissemination, and use of disaster related information serving as a sub-regional disaster information centre. It is a network of institutions across the Caribbean using ICTs to archive and retrieve data which is vital to their disaster preparedness planning. (Batchelor et al.)
Channel Africa, South Africa	Channel Africa broadcasts news from Africa via shortwave, satellite, and internet. Current affairs, economics, technology, education, environment, tourism, and sport are all covered by Channel Africa. (Op de Coul)
COPPADES, Nepal	COPPADES is an NGO in Nepal, which is trying to address the problem of lack of IT access and education in the rural areas of Nepal with focus on students and youth. (Op de Coul)
Deepalaya, India	Deepalaya is a chain of schools for slum children in India. They offer computer classes to their students, starting at the age of 10 (6th grade). Computer training is also included in their vocational programme. (Op de Coul)
DENIVA, Uganda	The project aims to make use of new technologies to facilitate vertical and horizontal integration of members. Areas of focus include information management, gender mainstreaming, environment, decentralized information exchange and others. (Batchelor et al.)
Digital Village, South Africa	Digital Village is an 'original' Telecentre. It was set up to be a resource for the a poor community to enable access to computers for training, information gathering, and communication. (Batchelor et al.)

Case study	Summary and author
eBrain (eBrain Forum of Zambia)	eBrain is a Zambian platform for information sharing, networking and lobbying in the field of ICT for development. eBrain has a web site and a quarterly newsletter and also organizes monthly offline meetings. (Op de Coul)
Entre Amigos, El Salvador	Entre Amigos ('Among friends') is an organization focusing on the human rights of homosexual persons. An internet connection, a web site and a cybercafé are the ICT instruments used to strengthen the gay community, to teach about their rights and about the prevention of HIV/AIDS. (Op de Coul)
Environmental Information Network, Ghana	The Environmental Information Network (EIN) Project of Ghana uses ICT to link the databases of two national environmental agencies. The database is publicly available for free use. Local and international researchers, government agencies and other environmental organizations can use its information to support decision-making, intervention strategies, and awareness campaigns about environmental protection, and they can also contribute to this knowledge pool. (Bridges.org)
Feminist International Radio Endeavour (FIRE), Costa Rica	FIRE is a women's Internet broadcaster, a radio station that transmits live events and conferences on specific dates. (Op de Coul)
FOOD, India	FOOD/IndiaShop aims to explore whether e-commerce can prove to be a source of income for women cooperatives and non-profits working in rural areas. Through this to a see also if they can train educated unemployed youth to function as e-marketers to promote products online and obtain a sustainable source of income for themselves. The project has explored using 'e-marketers' to set up a mechanism for ecommerce of handicrafts. (Batchelor et al.)
Global Voices, Kenya	Born from a strategic review process in Oxfam, this is a case of communities using video to increase awareness among government and their fellow community of the community needs. The objectives are to inform Oxfam in its strategic review process so that it could be more effective in alleviating poverty; and to give people a tool that would give them a voice so they could be heard expressing their concerns and possible solutions on issues that affect them. (Batchelor et al.)
Gyandoot, India	Gyandoot operates in Dhar District, a remote, tribal dominated, drought-prone area of Madhya Pradesh. The district has a population of 1.7 million, 54% of whom are tribal and 40% living below the poverty line. On 1 January 2000 Dhar District began the new millennium with the installation of a low-cost, self-sustainable rural intranet project, owned by the community. The name of the project, Gyandoot, literally means purveyor of knowledge. (Meera et al.) (Batchelor et al.)
iKisan project, India	iKisan is the ICT initiative of the Nagarjuna group of companies, the largest private entity supplying farmers' agricultural needs. iKisan was set up with two components, the iKisan.com web site, to provide agricultural information online, and technical centres at village level. The project operates in Andhra Pradesh and Tamil Nadu. However, it really proved popular in Andhra Pradesh where nine technical centres (kiosks) were established in different districts. Farmers are able to become members by paying Rs. 100 per year or Rs. 20 per month. Project services are available only to member farmers. (Meera et al.)

Case study	Summary and author
INCORPORE, Costa Rica	INCORPORE is an organization dedicated to the cultural sector in Central America. Its main objective is to contribute to the sustainability of Central American cultural expressions. (Op de Coul)
Indev, India	Indev is an initiative of the British Council to address problems faced by development managers in accessing development information on India. Indev uses ICTs and ItrainOnline to enhance their ICT training workshops. (Op de Coul)
Infocentros Association, El Salvador	Infocentros Association is dedicated to the democratisation of access to knowledge and of the creation, publication, and interchange of information. Forty telecentres and an Internet portal are the ICT tools Infocentros uses to achieve its objectives. (Op de Coul)
KCR (Kothmale Community Radio), Sri Lanka	Kothmale has an Internet Project through which information from the internet is used for their radio programmes, internet access is offered to the community and people are trained in the use of several ICT applications. (Op de Coul)
Kubatana, Zimbabwe	The Kubatana Project manages Kubatana.net, a web site portal that provides Zimbabwean civil society organizations with an online presence and a platform to voice their concerns about human rights abuses in their country. The project also offers courses that teach civil society organizations to use information communication technology (ICT) to further their goals. (Bridges.org)
KUMINFO, Ghana	KUMINFO aims to make data available and accessible to stakeholders involved in natural resource management. This is a GIS information gathering activity on a province wide scale. (Batchelor et al.)
MANAGE, India	MANAGE aims to increase rural farmers' access to information services. As an experiment in information extension, Manage has also set up a network of information kiosks. (Batchelor et al.)
MIGIS, China	MIGIS aims to make a significant contribution to the quality and effectiveness of participatory planning; by introducing the use of GIS and advanced graphic techniques into the PRA process; and, using the images produced in a way that would enhance the presentation and therefore the authority and impact of information collected in and provided by communities in which development intervention was planned. The project uses computer GIS systems and enhanced graphics to validate and present information gathered participatorily from illiterate and semi-literate communities. (Batchelor et al.)
NGO-CC (Non-Governmental Organisation's Co-ordinating Committee), Zambia	NGO-CC is establishing computer centres in the provinces of Zambia in order to give women access to the possibilities the internet offers. (Op de Coul)
Probidad Latin America, El Salvador	Probidad is a civic anticorruption organization. Its mission is 'to contribute to the eradication of corrupt practices in Latin America.' 80% of its work is done online using different websites and e-mail. (Op de Coul)
REVISTAZO, Honduras	REVISTAZO: To provide an alternative communication media source to inform the public about issues surrounding corruption and social injustice in order to promote Good Governance within Honduras. Using a web site, Revistazo is an online magazine that tackles sensitive political issues. (Batchelor et al.)

Case study	Summary and author
RLEK (Rural Litigation and Entitlement Kendra), India	RLEK has implemented a project that uses a wireless communication system to improve the exchange of information and communication between isolated tribal communities. (Op de Coul)
Technoclub/Paniamor Foundation, Costa Rica	Technoclub offers young people the opportunity to have access to new technologies with the objective of developing abilities and skills in the educational area as well as in work and social areas. (Op de Coul)
The SATELLIFE PDA Project, Africa	The SATELLIFE PDA Project explored questions related to the selection and design of appropriate, affordable technology and locally relevant content for use in the African healthcare environment, specifically targeted at assessing the usefulness of the PDA for (1) data collection and (2) information dissemination. Physicians, medical officers, and medical students tested the PDA in the context of their daily work environments in order to gain a perspective on the real issues that affect the adoption of technology. (Bridges.org)
The Tygerberg Children's Hospital and Rotary Telemedicine Project, South Africa	The Tygerberg Children's Hospital and Rotary Telemedicine Project in South Africa links specialists from Tygerberg Hospital to doctors at regional community or 'district' hospitals to improve healthcare in rural areas. The initiative has assembled its own telemedicine system using off-the-shelf computer equipment and software that is more affordable than commercial telemedicine systems. (Bridges.org)
UDS, Uganda	UDS: to help the poorest uplift themselves, in consultation with them, by providing appropriate information to facilitate development; communication to receive and distribute information; and training people in its practical applications. A small NGO has facilitated the setting up of centres which offers access to ICTs and training services for small business. (Batchelor et al.)
Video SEWA (Self-Employed Women's Association), India	Video SEWA shows that even an apparently sophisticated technology like video can be handled and used effectively by illiterate women to raise awareness and for lobby and advocacy. (Op de Coul)
Warana Wired Village Project, India	The Warana cooperative complex in Maharashtra has become famous as a fore-runner of successful integrated rural development emerging from the cooperative movement. The Warana cooperative sugar factory, registered in 1956, has led this movement, resulting in the formation of over 25 successful cooperative societies in the region. The total turnover of these societies exceeds Rs. 60 million. About 80% of the population is agriculture-based and an independent agricultural development department has been established by the cooperative society. The region is considered to be one of the most agriculturally prosperous in India. (Meera et al.)
Women'sNet, South Africa	Women'sNet offers a support and networking programme designed to enable (South) African women to use the internet to find the people, issues, resources, and tools needed for women's social activism. (Op de Coul)
YA-TV (Young Asia TV), Sri Lanka	YA-TV is a television network that covers issues of human rights, sustainable development, social justice, peace and conflict, information technology, and a host of other topics that impact on communities across Asia. (Op de Coul)

APPENDIX 3
Suggestions for further reading

ICT case stories

Batchelor, S., Norrish, P., Scott, N. and Webb, M. (2003) Eldis, *Sustainable ICT case histories*, http://www.sustainableicts.org

Cecchini, S. and Scott, C. (2003) UN, *Can Information and Communications Technology Applications Contribute to Poverty Reduction? Lessons from Rural India*, http://www.developmentgateway.org/download/181634/cecchini_scott_ICT.pdf

ICT Stories (Infodev and IICD) (2004) Infodev, *Keep in mind: overall analysis*, http://iconnect.osc.nl/stories/lbd/overall_analysis

Meera, Shaik A., Khamtani, A. and Rao, D.U.M. (2004) ODI, *Information and Communication Technology in Agricultural Development: A comparative analysis of three projects from India*, http://www.odi.org.uk/agren/papers/agrenpaper_135.pdf

Op de Coul, M. (2003) Bellanet, *ICT for development Case Studies*, http://www.bellanet.org/leap/docs/BDOsynthesis.pdf

Raab, R.T., Woods, J. and Abdon, B.R. (2003) Asia Pacific Regional Technology Centre (APRTC), Thailand, *The role of eLearning in promoting sustainable agricultural development in the GMS: educating knowledge intermediaries*, http://www.ait.ac.th/digital_gms/Proceedings/D32_ROBERT_T_RAAB.pdf

ICTs and development

Adeya, C.N. (2002) IDRC, *ICTs and Poverty: A literature review*, http://web.idrc.ca/uploads/userS/10541291550ICTPovertyBiblio.doc

Badshah, A., Khan, S. and Garrido, M. (2004) UN, *Connected for Development: Information Kiosks and Sustainability*, http://www.unicttaskforce.org/wsis/publications/ Connected%20for%20Development.pdf

Ballantyne, P. (2002) IICD, *Collecting and propagating local development content*, http://www.ftpiicd.org/files/research/reports/report7.pdf

Bridges.org (2001), *Spanning the Digital Divide*, http://www.bridges.org/spanning/report.html

Bridges.org (2004) *The Eight Habits of Highly Effective ICT-Enabled Development Initiatives*, http://www.bridges.org/our_approach/eight_habits.html

Chapman, R. and Slaymaker, T. (2002) ODI, *ICTs and Rural Development: Review of the Literature, Current Interventions and Opportunities for Action*, http://www.odi.org.uk/publications/working_papers/wp192.pdf

Gumicio Dagron, A. (2001) The Rockefeller Foundation, *Making Waves: Stories of Participatory Communication for Social Change*, http://www.rockfound.org/Documents/421/makingwaves.pdf

Michiels, S.I. and Van Crowder, L. (2001) FAO, *Discovering the 'Magic Box': local appropriation of information and communication technologies (ICTs)*, http://www.fao.org/sd/2001/KN0602a_en.htm

O'Farrell, C., Norrish, P. and Scott, A. (1999) AERDD, *Information and Communication Technologies (ICTs) for Sustainable Livelihoods*, http://www.rdg.ac.uk/AcaDepts/ea/AERDD/ICTBriefDoc.pdf

Stoll, K., Menou, M.J., Camacho, K. and Khellady, Y. (2001) IDRC, *Learning about ICTs' role in development: A framework toward a participatory, transparent and continuous process*, http://www.bellanet.org/leap/docs/evaltica.doc?OutsideInServer=no

Paisley, L. and Richardson, D. (1998) FAO, *Why the First Mile and not the Last?*, http://www.fao.org/docrep/x0295e/x0295e03.htm

Donor policy on knowledge sharing

DFID (1997) *Eliminating World Poverty: A Challenge for the 21st Century*, http://www.dfid.gov.uk/PolicieAndPriorities/files/whitepaper1997.pdf

Marker, P., McNamara K. and Wallace, L. (2002) DfID, *The significance of information and communication technologies for reducing poverty*, http://www.dfid.gov.uk/Pubs/files/ict_poverty.pdf

World Bank (1998) *World Bank World Development Report 1998/99: Knowledge for Development*, http://www.worldbank.org/wdr/wdr98/index.htm

ITDG's research into ICTs

Aley, R. (2003) ITDG, *Pro-poor Satellite Broadcasting: Reality or Myth?* [unpublished work]

Aley, R., Waudo, A. and Muchiri, L. (2004) ITDG, *MicroMedia card pack.*

Lloyd Laney, M. (2003) ITDG, *Making knowledge networks work for the poor*, http://www.itcltd.com/docs/mknwp%20project%20final%20report.pdf

Schilderman, T. (2002) ITDG, *Strengthening the knowledge and information systems of the urban poor*, http://www.itdg.org/html/shelter/docs/kis_urban_poor_report_march2002.doc

Wakelin, O. (2003) ITC, *ITC Demand Assessment Exercise for Small Business Information*, http://www.itcltd.com/icts/

References

Adeya, C.N. 2002 IDRC *ICTs and Poverty: A literature review* [Online].
   Available at http://web.idrc.ca/uploads/user-S/10541291550ICTPovertyBiblio.doc
   [Accessed 2-7-2004]

Aley, R. 2003a, *Pro-poor Satellite Broadcasting: Reality or Myth?* [Unpublished work]

Aley, R. 2003b, 'Rural information systems – can technology help?', *BASIN News* no. 25.

Aley, R., Waudo, A., & Muchiri, L. 2004, *MicroMedia card pack* ITDG

Badshah, A., Khan, S., and Garrido, M. 2004 *UN Connected for Development: Information Kiosks and Sustainability* [Online].
   Available at
   http://www.unicttaskforce.org/wsis/publications/ConnectedforDevelopment.pdf
   [Accessed 20-5-2004]

Ballantyne, P. 2002 IICD *Collecting and propagating local development content* [Online].
   Available at http://www.ftpiicd.org/files/research/reports/report7.pdf
   [Accessed 20-9-2003]

Ballantyne, P., Labelle, R. and Rudgard, S. 2000 ECDPM *Information and Knowledge Management: Challenges for Capacity Builders* [Online].
   Available at http://www.chs.ubc.ca/lprv/PDF/lprv0075.pdf
   [Accessed 17-11-2004]

Barnard, G. 2003 Institute of Development Studies, Brighton *Knowledge Sharing in Development Agencies: Knowledge Fortress or Knowledge Pool?* [Online].
   Available at http://www.ids.ac.uk/ids/info/eadi/
   [Accessed 29-9-2003]

Batchelor, S., Norrish, P., Webb, M., and Scott, N. 2003 Gamos *Sustainable ICT case histories* [Online].
   Available at http://www.sustainableicts.org
   [Accessed 11-11-2003]

Batchelor, S. and Sugden, S. 2003 Gamos *An Analysis of Infodev Case studies: Lessons Learned* [Online].
   Available at http://www.sustainableicts.org/infodev/infodevreport.pdf
   [Accessed 31-1-2005]

Beardon, H. 2004, *ICT for development: empowerment or exploitation? Learning from the Reflect ICTs project*, ActionAid.

Besemer, H., Addison, C. and Ferguson, J. 2003 DfID *Fertile ground : opportunities for greater coherence in agricultural information systems* [Online].
   Available at http://www.ftpiicd.org/files/research/reports/report19.pdf
   [Accessed 20-08-2004]

Best, M.M. and Maclay, C.M. 2002 Centre for International Development at Harvard University *Community Internet Access in Rural Areas: Solving the Economic Sustainability Puzzle* [Online].
   Available at http://www.cid.harvard.edu/cr/pdf/gitrr2002_ch08.pdf
   [Accessed 19-1-2005]

Boyle, G. 2002b, 'Putting context into ICTs in international development: an institutional networking project in Vietnam', *Journal of International Development*, vol. 14, no. 1, pp. 101–12.

Bridges.org 2001 *Spanning the Digital Divide* [Online].
Available at http://www.bridges.org/spanning/report.html
[Accessed 22-10-2003]

Bridges.org 2003 *Evaluation of the SATELLIFE PDA Project, 2002 – Testing the use of handheld computers for healthcare in Ghana, Uganda, and Kenya* [Online].
Available at http://www.bridges.org/satellife/
[Accessed 29-11-2004]

Bridges.org 2004 *The 8 Habits of Highly Effective ICT-Enabled Development Initiatives* [Online].
Available at http://www.bridges.org/our_approach/eight_habits.html
[Accessed 26-7-2004]

Carlsson, C. 2002 MANDE *How can we learn more from what we do? Evaluation and evidence-based communications for development* [Online].
Available at http://www.mande.co.uk/docs/Summary%20record-FINAL.doc
[Accessed 20-10-2003]

Carman, B. 2005 IDRC *Information and Communication Technologies for Development in Africa: Volume 1 Opportunities and Challenges for Community Development* [Online].
Available at http://web.idrc.ca/en/ev-33000-201-1-DO_TOPIC.html
[Accessed 27-1-2005]

Cecchini, S. and Scott, C. 2003 UN *Can Information and Communications Technology Applications Contribute to Poverty Reduction? Lessons from Rural India* [Online].
Available at
http://www.developmentgateway.org/download/181634/cecchini_scott_ICT.pdf
[Accessed 22-6-2004]

Chapman, R., Slaymaker, T., & Young, J. 2003, *Livelihoods Approaches to Information Communication in Support of Rural Poverty Elimination and Food Security*, Overseas Development Institute.

Crewe, E. and Young, J. 2002 ODI *Bridging Research and Policy: Context, Evidence and Links* [Online].
Available at http://www.odi.org.uk/publications/working_papers/wp173.pdf
[Accessed 20-7-2004]

Davenport, T. & Prusak, L. 1998, *Working Knowledge*, Harvard Business School Press.

Devlin, K. 1999, *Infosense: Turning Information into Knowledge*, W.H. Freeman: New York.

Fisher, J., Odhiambo, F. & Cotton, A. 2003, *Spreading the Word Further*, WEDC, Loughborough University.

Flor, A.G. 2001 World Bank Publications *ICT and poverty: the indisputable link* [Online].
Available at
http://www.worldbank.org/html/extdr/offrep/eap/eapprem/infoalexan.pdf
[Accessed 20-08-2004]

Gigler, B.S. 2004 LSE *Including the Excluded- Can ICTs empower poor communities? Towards an alternative evaluation framework based on the capability approach* [Online].
Available at http://cfs.unipv.it/ca2004/papers/gigler.pdf
[Accessed 13-12-2004]

Gomez, R. and Casadiego, B. 2002 IDRC *Carta a la Tía Ofelia: Siete propuestas para un desarrollo equitativo con el uso de Nuevas Tecnologías de Información y Comunicación* [Online].
Available at http://www.idrc.ca/pan/ricardo/publications/Ofelia.htm
[Accessed 25-11-2004]

Gumicio Dagron, A. 2001 The Rockefeller Foundation, *Making Waves: Stories of Participatory Communication for Social Change* [Online].
Available at http://www.rockfound.org/Documents/421/makingwaves.pdf
[Accessed 17-10-2003]

Hagen, I. 2005 IICD *Going beyond a project approach: embedding ICT support in a wider development context* [Online].
Available at
http://www.capacity.org/Web_Capacity/Web/UK_Content/Download.nsf/0/7102
14ABB27990FEC1256F5D00424A8D/$FILE/Capacity_23_ENG_FINAL.pdf
[Accessed 27-1-2005]

Hammond, B. 2001 OECD [Online].
Available at
http://www1.oecd.org/dac/digitalforum/docs/DO_IssuesConclusions_En.pdf
[Accessed 22-10-2003]

Harris, R. 2004 APDIP *ICTs for Poverty Alleviation* [Online].
Available at http://eprimers.apdip.net/series/info-economy/poverty.pdf
[Accessed 9-12-2004]

Heeks, R. 2002, 'i-development not e-development: special issue on ICTs and development', *Journal of International Development*, vol. 14, no. 1, p. 1.

Hovland, I. 2003a ODI *Communication of research for poverty reduction: a literature review* [Online].
Available at http://www.odi.org.uk/publications/working_papers/wp227.pdf
[Accessed 20-2-2004]

Hovland, I. 2003b ODI *Knowledge Management and organisational learning: An international development perspective* [Online].
Available at
http://www.odi.org.uk/RAPID/Bibliographies/KM/KM_Review_00.html
[Accessed 10-9-2003]

Infodev 2004 *Keep in mind: overall analysis* [Online].
Available at http://iconnect.osc.nl/stories/lbd/overall_analysis
[Accessed 29-7-2004]

ITDG 2003 ITDG *Women's Voices* [Online].
Available at http://ww.itdg.org/?id=womens_voices
[Accessed 29-11-2004]

King, K. and McGrath, S. 26-8-2003 Centre of African Studies, University of Edinburgh *Knowledge sharing in development agencies: Lessons from Four Cases* [Online].
Available at www.ed.ac.uk/centas/fgpapers.html
[Accessed 28-8-0003]

Lloyd Laney, M. 2003a CIMRC *Advocacy Impact Assessment Guidelines* [Online].
Available at http://www.cimrc.info/pdf/news/Impactassess.pdf
[Accessed 29-11-2004]

Lloyd Laney, M. 2003b Infrastructure Connect *Making Information User-Driven* [Online].
Available at http://www.infrastructureconnect.info/pdf/news/Userdriveninfo.pdf
[Accessed 20-4-2003]

Lloyd Laney, M. 2003c ITDG *Making knowledge networks work for the poor* [Online].
Available at http://www.itcltd.com/docs/mknwp project final report.pdf
[Accessed 10-7-2003]

Ludin, J. 2003 BOND *Where are we. . . with North-South Learning?* [Online].
Available at http://www.bond.org.uk/networker/2002/dec02/opinion.htm
[Accessed 20-3-2003]

Meera, S.A., Khamtani, A. and Rao, D. U. M. 2004 ODI *Information and Communication Technology in Agricultural Development: A comparative analysis of three projects from India* [Online].

Available at http://www.odi.org.uk/agren/papers/agrenpaper_135.pdf
[Accessed 18-5-2004]

Meyer, C. A. 1997, 'The Political Economy of NGOs and Information Sharing', *World Development*, vol. 25, no. 7, pp. 1127–1140.

Michiels, S.I. and Van Crowder, L. 2001 FAO Discovering the 'Magic Box': local appropriation of information and communication technologies (ICTs) [Online].
Available at http://www.fao.org/sd/2001/KN0602a_en.htm
[Accessed 14-7-2004]

Nicol, C. 2003 APC ICT Policy Handbook [Online].
Available at http://www.apc.org/english/rights/handbook/index.shtml
[Accessed 13-1-2005]

Nonaka, I. & Takeuchi, H. 1995, *The Knowledge-Creating Company*, Oxford University Press.

O'Farrell, C., Norrish, P., and Scott, A. 1999 AERDD *Information and Communication Technologies (ICTs) for Sustainable Livelihoods* [Online].
Available at http://www.rdg.ac.uk/AcaDepts/ea/AERDD/ICTBriefDoc.pdf
[Accessed 26-7-2004]

Op de Coul, M. 2003 Bellanet *ICT for development Case Studies* [Online].
Available at http://www.bellanet.org/leap/docs/BDOsynthesis.pdf
[Accessed 18-5-2004]

Orlikowski, W. J. 2000, 'Using technology and constituting structures: A practice lens for studying technology in organizations', *Organization Science*, vol. 11, no. 4, p. 404.

Paisley, L. and Richardson, D. 1998 FAO *Why the First Mile and not the Last?* [Online].
Available at http://www.fao.org/docrep/x0295e/x0295e03.htm
[Accessed 26-7-2004]

Prahalad, C. K. & Hammond, A. 2002, 'Serving the world's poor, profitably', *Harvard Business Review*, vol. 80, no. 9, p. 48.

Primo Braga, C. A., Daly, J. A., & Sareen, B. 2003, 'The Future of Information and Communication Technologies for Development', pp. 1–17.

Raab, R.T., Woods, J. and Abdon, B.R. 2003 Asia Pacific Regional Technology Centre (APRTC), Thailand *The role of eLearning in promoting sustainable agricultural development in the GMS: educating knowledge intermediaries* [Online].
Available at
http://www.ait.ac.th/digital_gms/Proceedings/D32_ROBERT_T_RAAB.pdf
[Accessed 11-7-2004]

Schech, S. 2002, 'Wired for change: the links between ICTs and development discourses', *Journal of International Development*, vol. 14, pp. 13–23.

Schilderman, T. 2002 ITDG *Strengthening the knowledge and information systems of the urban poor* [Online].
Available at
http://www.itdg.org/html/shelter/docs/kis_urban_poor_report_march2002.doc
[Accessed 10-10-2003]

Stiglitz, J. 1999 Global Development Network *Scan Globally, Reinvent Locally: Knowledge Infrastructure and the Localisation of Knowledge* [Online]
Available at: http://www.gdnet.org/pdf/226_GDNfinal.pdf
[Accessed 25-8-2003]

Stoll, K., Menou, M.J., Camacho, K.and Khellady, Y. 2001 IDRC *Learning about ICT's role in development: A framework toward a participatory, transparent and continuous process.* [Online].
Available at http://www.bellanet.org/leap/docs/evaltica.doc?OutsideInServer=no
[Accessed 8-8-2004]

Tschang, T., Chuladul, M., & Le, T. T. 2002, 'Scaling-up information services for development: a framework of increasing returns for telecentres', *Journal of International Development*, vol. 14, no. 1, pp. 129–141.

van der Velden, M. 2002, 'Knowledge facts, knowledge fiction: the role of ICTs in knowledge management for development', *Journal of International Development*, vol. 14, pp. 25–37.

Waisbord, S. 2003 Communications Initiative Family tree of theories, methodologies and strategies in development communication [Online].
Available online at http://www.gdnet.org/pdf/226_GDNfinal.pdf
[Accessed 11-11-2003]

Weigel, G. and Waldburger, D. 2004 SDC / GKP *ICT4D – Connecting People for a Better World* [Online].
Available at http://www.globalknowledge.org/ict4d/index.cfm?menuid=43
[Accessed 18-11-2004]

World Bank 1998a World Bank *Indigenous knowledge for development: a framework for action* [Online].
Available at http://www.worldbank.org/afr/ik/ikrept.pdf
[Accessed 8-2-2005]

World Bank 1998b World Bank *World Development Report 1998/99: Knowledge for Development* [Online].
Available at http://www.worldbank.org/wdr/wdr98/index.htm
[Accessed 8-2-2003]

World Bank 2001 *World Development Report 2000/2001: Attacking Poverty* [Online].
Available at http://www.worldbank.org/poverty/wdrpoverty/report/index.htm
[Accessed 20-8-2004]

www.ingramcontent.com/pod-product-compliance
Lightning Source LLC
La Vergne TN
LVHW062317070326
832902LV00029B/4644